Understanding and Teaching Primary Mathematics in Australia

Written by experienced teacher educator and author, Tony Cotton, and two Australian primary teachers, Jess Greenbaum and Michael Minas, *Understanding and Teaching Primary Mathematics in Australia* combines pedagogy and mathematics subject knowledge to build teachers' confidence both in their mathematical subject knowledge and in their ability to teach mathematics effectively.

The book covers all the key areas of the Australian National Curriculum for mathematics from teaching number and calculation strategies to exploring geometry and statistics. There are also chapters that deal with the teaching of mathematics in the Early Years, inclusive approaches to mathematics teaching and teaching mathematics using ICT.

Stimulating, accessible and containing a wealth of practical ideas you can use in your classroom, *Understanding and Teaching Primary Mathematics in Australia* is an essential purchase for graduate and practicing teachers alike.

Tony Cotton was previously Associate Dean and Head of Education at Leeds Metropolitan University, UK. Since 2012, he has been working as a freelance writer and education consultant. He has written widely for mathematics teachers and writes pupil textbooks. Tony has worked internationally developing curricula and running teacher workshops. He currently edits *Mathematics Teaching*, the journal of the Association of Teachers of Mathematics.

Jess Greenbaum is a primary school mathematics learning specialist in Melbourne, Australia.

Michael Minas is a Melbourne-based education consultant who specialises in working with primary schools in the area of mathematics.

Understanding and Teaching Primary Mathematics in Australia

1ST AUSTRALIAN EDITION

Tony Cotton, Jess Greenbaum
and Michael Minas

Routledge
Taylor & Francis Group

LONDON AND NEW YORK

Cover image: Getty Images

First published 2023
by Routledge
4 Park Square, Milton Park, Abingdon, Oxon OX14 4RN

and by Routledge
605 Third Avenue, New York, NY 10158

Routledge is an imprint of the Taylor & Francis Group, an informa business

British Library Cataloguing-in-Publication Data
A catalogue record for this book is available from the British Library

Library of Congress Cataloging-in-Publication Data
A catalog record has been requested for this book

ISBN: 978-1-032-32463-0 (hbk)
ISBN: 978-1-032-32462-3 (pbk)
ISBN: 978-1-003-31515-5 (ebk)

DOI: 10.4324/9781003315155

Typeset in Frutiger
by Apex CoVantage, LLC

Access the companion website: www.routledge.com/cw/cotton

CONTENTS

Chapter 6   Number: Calculation (addition, subtraction, multiplication and division)   131

Chapter 7   Patterns and algebra   172

Contents

Chapter 11 Teaching and learning mathematics in the Early Years 267

Chapter 12 Issues of inclusion 297

PREFACE TO THE AUSTRALIAN EDITION

Introduction

Welcome to the Australian edition of *Understanding and Teaching Primary Mathematics*. This edition has been reworked from the popular 4th edition of the book in the UK. This ensures the book is of direct relevance to educators in Australia. Tony has been lucky enough to work in over 30 countries and it is interesting to see how similar the mathematics curricula are in these countries and how the concerns that early career teachers have around their own confidence in mathematics are shared worldwide. Jess and Michael are both experienced teachers from Melbourne. Jess teaches in a primary school and is the mathematics learning specialist there. Michael is an education consultant who specialises in working with primary schools in the area of mathematics. This places them in the perfect position to work with Tony on developing the book for an Australian readership.

The particular approach of this book is to develop your mathematical subject knowledge through exploring the learning and teaching of mathematics. The book will support you in developing your own understanding of mathematics by examining the misconceptions of the learners you will work with and by developing your own repertoire of teaching strategies so you can immediately see the impact of your own learning on your teaching practice.

The *Australian Curriculum: Mathematics*, suggests that effective teaching of mathematics should ensure that students:

> are confident, creative users and communicators of mathematics, able to investigate represent and interpret situations in their personal and work lives and as active citizens.

This aim is not something that is unique to Australia. The Primary Years Programme followed by learners in over 100 countries around the world is based around cross-curricular themes and "structured, purposeful inquiry that engages students actively in their own learning." Teachers are encouraged to draw on learners' prior knowledge, build on this knowledge through new experiences and provide opportunities for reflection and consolidation. (See www.ibo.org/en/programmes/primary-years-programme for more information.)

The *Australian Curriculum: Mathematics* also suggests that the rationale for learning and teaching mathematics is that it "creates opportunities for and enriches the lives of Australians." For this to take place, we would suggest that you, as a teacher, need to feel that your own life has been enriched by mathematics. So, one aim for this book is that it will enrich you mathematically. We hope this book will support you in becoming a high-quality mathematics teacher in this sense.

The aim of the book

The book aims to support you in developing the three key areas of mathematical subject knowledge. These are

- **Mathematical knowledge:** the book will support you in developing your own understanding of mathematics through engaging you in activities and investigations.
- **Curriculum knowledge:** through reading the book you will come to understand which areas of mathematics are expected to be taught at each stage of learning and understand the progression within learning and teaching mathematics, so you will know which areas of mathematics your learners should have encountered and where they will go next.
- **Pedagogical knowledge:** the book draws on examples from the classroom so that you can see the best ways to introduce your learners to particular mathematical ideas.

The book is practical and models good primary practice in the way it supports you in developing your own knowledge. In this way you will both develop your own understanding and your teaching repertoire.

Who should use this book?

This book is aimed at all those who wish to develop their own mathematical knowledge in order to improve their confidence in teaching mathematics:

Pre-service teachers

Pre-service teachers will find this book can be used throughout their course to develop a personal portfolio and to support them on their school placements. They will find it useful for assignments which look at particular areas of mathematics or to give ideas for teaching on placement.

Graduate teachers

Graduate teachers will be able to draw on the book to support them in areas they have not previously taught or that were not covered specifically during their training.

Experienced teachers

Experienced teachers will find this book useful when visiting areas of mathematics new to them or when working with new year levels. They may also find the book useful to refresh their understanding of areas they have not taught for some time.

Distinctive features

Big ideas

Each section starts with the big ideas that underpin the particular strand of mathematics covered. This allows you to see the big picture immediately and understand how the mathematics you are teaching falls within the overall landscape of the curriculum.

Progression

Each chapter gives you an overview of progression across the Early Years and throughout the primary years so that you understand which mathematical ideas are appropriate to the children you are teaching. It also helps you understand what knowledge your students will bring with them and which mathematical ideas they should progress on to next. This is structured so that you are introduced to the foundations for learning in each area of mathematics and then the ideas which support learners in beginning to understand this area of mathematics are described. The skills and understandings that students should be taught to become confident mathematicians are outlined, and finally suggestions for extending those learners in your classroom who are skilled mathematicians are described.

Links to the classroom and research

Each chapter makes links to key pieces of research and curriculum development. This will support you in writing assignments and in seeing direct links to the classroom. There are also activities that you can try out in your classroom and ideas to support you in developing your own resources.

Audit and portfolio tasks

These tasks allow you to build a portfolio of evidence to show that you have acquired the subject knowledge you need to gain qualified teacher status.

Teaching points

These draw on common misconceptions in learners and support you in knowing how you can deal with these misconceptions in the classroom. As mentioned earlier in this preface, it is important that you see your learners' misconceptions as teaching points. By understanding what the most common misconceptions are likely to be you can prepare to overcome learners' common difficulties.

Classroom activities

You will also find boxes in the text that contain activities you can carry out with the learners in your classroom. These are designed to support you in developing your own mathematical knowledge through teaching particular concepts. We always find that we become much more secure in our knowledge if we have to teach something. There is a companion volume to the book called *Teaching for Mathematical Understanding: Practical ideas for outstanding primary lessons*. This book contains all these activities with full lesson plans to help you try them out successfully.

Assessment

Assessing is a key part of learning. It helps you know what your students currently understand so that you can plan for the next piece of learning. It also helps you understand how effective your teaching has been and to uncover any areas of difficulty that you need to revisit. Each chapter offers you an assessment activity to use with the students you are teaching.

Case studies

Each chapter contains a case study that gives you a lesson plan and an evaluation for each strand of mathematics. This makes a direct link between your own subject knowledge and your teaching.

Drawing on our shared experience of teaching mathematics in Australia, the UK and many countries around the world, we hope to have written a book that will support you in developing your subject knowledge. We do this through encouraging you to reflect on your own mathematical understandings and by exploring your teaching of mathematics. Our aim would be that through using this book you will develop your subject knowledge and your teaching so that you can confidently and skilfully teach any area of mathematics to learners of all ages.

Guided Tour

Chapter introductions and starting points include a mathematical problem or piece of research to illustrate a concept or theme that is built upon within the chapter.

Each chapter provides an overview of **progression** across the primary years to aid understanding of which mathematical ideas are appropriate to the children you are teaching.

Big ideas deal with the fundamental ideas which underpin each particular strand of mathematics covered. This allows you to see the big picture immediately and understand how the different strands knit together.

Self-audit and portfolio tasks allow you to build a portfolio of evidence to show that you have acquired the subject knowledge you need to gain qualified teacher status.

Teaching points highlight children's common mathematical misconceptions.

Taking it further boxes make links to key pieces of research and curriculum development, to support you in writing assignments and in seeing direct links to the classroom.

The **In practice** section is a case study which gives you a lesson plan and an evaluation for each strand of mathematics. This makes a direct link between your own subject knowledge and your teaching.

Key terms that are emboldened throughout the text can be found in the glossary at the end of the book.

ACKNOWLEDGMENTS

Tony

The most important people to me in terms of teaching me about mathematics education are Felix and Tate, two of my grandsons. Aged nine and four, they still revel in learning how to see the world mathematically and indulge their grandad in letting him learn alongside them. They quickly let me know if activities I plan are appropriate or not. You can be confident that everything in this book has been approved by them.

I also want to thank my three children Holly, Adam and Sam and my patient and inspirational life partner Helen. Helen and I have always worked together since we first met teaching in adjacent rooms. That was in 1986 – we have learned a lot since then but are still both inspired by all the teachers and graduate teachers we work with.

Jess

Thank you, Kazz for supporting me in every adventure, no matter the shape or size. Michael, the inspiration and opportunity you've shared with me and continue to share is invaluable.

Michael

Carla, thanks for putting up with me through the process of working on another book. And Nash, Isaiah and Genevieve, thanks for being the most beautiful trio of kids, we are truly blessed to have you in our lives.

CHAPTER 1

TEACHING AND LEARNING PRIMARY MATHEMATICS

Introduction

Learning maths was like a dagger going through my skull – no matter how hard I tried it wasn't good enough. This was really frustrating 'cause I knew there were really important things to discover and when I'm teaching and the kids really get it it's like they've found diamonds. I don't know whether it was my fault or the teacher's.

This is the way a student of Tony's shared the experience of being both a learner and a teacher of mathematics. He had asked a group to describe their memories of learning mathematics at primary school. It vividly details the concerns that some of you may have that you do not have sufficient mathematical understanding to be as effective a teacher as you want to be. It also describes the joys that come with seeing that you have been able to explain a complex mathematical concept well. During the same activity another student remembered how a feeling of panic can set in if you feel as though you have not got a good grasp of the area that you are teaching.

My problem is being put on the spot – I can't think then, lots of little warnings run through my head the moment I'm asked a question.

Some of you will have had very positive experiences of learning mathematics and some of you may have felt as though you could not understand what was being asked of you. Whichever is true for you, working at your own subject knowledge will support you in feeling much more able to find those diamonds that are lurking and to feel confident that you will be able to answer whatever question a student throws at you.

DOI: 10.4324/9781003315155-1

Research projects such as the one described above suggest that the best way for someone to support teachers in developing mathematical subject knowledge is by engaging them in actively questioning and reflecting on their learning. Some beginning teachers do not see themselves as "good" at mathematics because of the way in which they have experienced mathematics as a learner. Through exploring mathematics in a new way, they gradually change their view of what being "good at mathematics" is and can begin to see themselves as mathematicians. Introducing their students to a form of mathematics that they have recently experienced themselves allows them to be both tentative in terms of the possible outcomes whilst being secure in the process. They know, or they trust, that interesting things will happen in terms of children learning mathematics. And they notice that through exploring mathematics in an open and investigatory way their learners are themselves developing as young mathematicians. These beginning teachers are moving towards a belief that mathematics is about questioning, exploring and justification.

The beginning of this process is to understand what we might mean by subject knowledge. This chapter offers a definition of subject knowledge for teaching mathematics, as well as setting the scene for the rest of the book. It describes how working with the book will support you in developing your own mathematical subject knowledge. Armed with this understanding you can plan how best you might use the book.

Starting point

Perhaps we can use the following example to exemplify our view of mathematical subject knowledge? A pre-service teacher had placed a multiplication grid (you fill in the blanks by multiplying together the numbers at the end of the row and column) on the whiteboard for the students in the class to complete as the warm-up activity for a mathematics lesson.

As the multiplication grid was revealed there were sounds of complaint from the students. One said to the pre-service teacher "You don't put fractions in a number grid and we only go up to 10!" The pre-service teacher patiently persuaded the students to accept this "new" version of what had become an everyday activity for the class and then asked them to complete the grid.

After ten minutes the pre-service teacher stopped the class and asked, "Which column did you fill in first?" One of the students put their hand up and said "48." The pre-service teacher paused and said, "I don't want you to tell me any answers. I want you to tell me which column you filled in first." At first the students did not understand why the answer for the top-left square wasn't seen as important. Then realisation struck. One student said, "You could do the twos first. That would be really easy."

×	8	2	$\frac{1}{2}$	5	$\frac{1}{4}$	10	4
6							
3							
7							
9							
4							
5							
12							

Another suggested starting with the tens for the same reason. By the end of the discussion, the class had come to understand that mathematical thinking was about looking carefully at a problem and finding the most effective or efficient way of solving it rather than simply following a process that had worked previously. So, for this grid, if you complete the twos column you can then complete the fours and eights columns, by doubling. Similarly, filling in the tens column allows you to complete the fives column by halving. The same process will sort out the half and quarter columns.

Maybe you can try this activity out with some of your friends or, even better, with some of your students. We would bet that most people will start at the top left corner and work systematically through the grid. This shows how deeply engrained the feeling is that there is a "right-way" to complete the activity. If you suggest an alternative, the students you are working with might say that it, "feels like cheating." It is hard for them to break away from the image of learning mathematics that they carry with them from previous experiences at school. So, if you do not feel confident in your own mathematics knowledge this is, in part, due to your experiences as a learner of mathematics. We hope this book will allow you to make a break from your prior experience and realise that you can become a confident mathematician.

Why is mathematical subject knowledge important?

Whether you are training to become a teacher or are an experienced teacher who wants to improve their mathematical subject knowledge, the process is the same. Someone with good mathematical subject knowledge is able to be confident in what they are teaching and, more importantly, will be seen as confident by their students. We all learn better from someone who we believe is both confident in their own knowledge and who is passionate about sharing that knowledge with us.

That is why the focus of this book is on developing your mathematical subject knowledge by sharing activities that you can use in the classroom. We hope that by engaging you in mathematics that you see as relevant to your teaching, you can become as passionate as us about mathematics itself. This will ensure that the students you teach will learn well. We will have been successful in writing this book if you as a teacher and your students as learners enjoy your mathematics and if you become more confident learners of mathematics together.

We would argue that there is a direct link between good mathematical subject knowledge and effective teaching and learning of mathematics. We would also suggest that there is a direct connection between having good subject knowledge and being able to make appropriate choices about the way in which to explore particular mathematical ideas. Before moving on, let us define exactly what we mean by "good mathematical subject knowledge."

What is good mathematical subject knowledge?

Research on teaching suggests that teachers draw on three forms of knowledge in order to teach effectively. The first is knowledge of the subject itself; you need to feel confident in your own mathematical subject knowledge in order to be able to teach effectively. Teachers also need an understanding of the curriculum they are expected to teach, so you need to be clear which mathematical ideas and concepts are appropriate to the age range of the students you are teaching. Finally, you need to understand which are the most appropriate strategies and tasks to engage your students in learning a particular mathematical idea. In order to develop your skills in teaching and learning primary mathematics this book must focus on these three areas. The book aims to do this in the following ways:

Mathematical subject matter knowledge

The tasks within this book ask you to explore mathematical ideas through engaging in investigations and practical activities. Our aim here is that you develop what is called a relational understanding of mathematics rather than an instrumental understanding. In an article first published in *Mathematics Teaching*, the journal of the Association of Teachers of Mathematics in the UK, in 1976, Richard

Skemp introduced this idea, and it has been drawn on, around the world, ever since. Relational understanding means that you move beyond a mechanical or rote view of mathematical processes in order to see and understand the links and connections between the different areas of mathematics. You will not be forced to try to "remember" mathematical rules and processes that you were taught at school, instead you will come to an understanding of how these processes actually work.

Mathematics curriculum knowledge

The book supports you in developing an understanding of the progressive nature of coming to understand the different areas of mathematics. Each chapter opens with an outline of the key concepts that students will be taught. These are divided into three groups:

1   Beginning to understand
2   Becoming confident
3   Extending learning.

We believe that we can use a student's understanding as a starting point for development whatever their prior attainment or experience. Having an understanding of the progression of mathematical skills and concepts within a strand allows you to make decisions about what is appropriate for your students at their stage of learning. Through careful assessment you can understand the stage that your students have reached. Your curriculum knowledge then allows you to plan carefully for the next stage of learning. You will also become aware of the ideas that your students will meet when they move on from your classroom. It is important for us to be clear about the next stages of learning both so that we can extend individuals' learning and so that we can feel comfortable that our students are ready for the next stage.

Pedagogical content knowledge

When you teach any area of mathematics you make choices: you choose a particular teaching approach; you may choose examples to introduce a concept; you may choose a particular way to group your learners and you choose tasks for your students to work on. A teacher with good pedagogical content knowledge has an awareness of the choices that they made and why they made those particular choices. This area of subject knowledge develops as you become more experienced – then you are able to draw on an ever-growing bank of tasks and strategies. This book allows you to share other teachers' experiences through drawing on a wide range of tasks and strategies and offering you a reflective commentary on the planning and teaching process.

To summarise, a teacher with good subject knowledge has the following understandings:

- mathematical subject matter knowledge (they understand the mathematics)
- curriculum subject knowledge (they know the requirements of the curriculum)
- pedagogical subject knowledge (they can make good choices to help them teach the mathematics).

What makes a good teacher of mathematics?

You probably have in mind a mathematics teacher that you learnt from at school whom you see as a good teacher. The fact that you remember them as a good teacher of mathematics probably means that they had good subject knowledge drawing on the three areas described previously. We would suggest that this meant they had some or all of the following skills.

They could anticipate possible misconceptions or misunderstandings and knew how best to support learners at coming to new, more effective conceptions. (Some examples are given later.) This knowledge does not just come through experience. One of the aims of this book is to share common misconceptions with you so that you can plan ahead to support your learners. Knowledge of common misconceptions allows you to respond flexibly and appropriately to the difficulties that children experience.

Examples of such misconceptions might be:

- you can't divide smaller numbers by larger ones
- division always makes numbers smaller
- the more **digits** a number has, the larger its value
- all shapes with larger **areas** have larger **perimeters**.

You may want to take a moment to think of counter examples to these statements. (For example, $5 \div 10 = 0.5$; $15 \div 0.5 = 30$; 0.000526 is smaller than 2; for the shape example try exploring different **rectangles** with a perimeter of 12cm.)

Teachers with good subject knowledge can also generate probing questions that allow children to articulate their current understandings and through this articulation come to better developed understandings. Often a probing question simply asks how a student has worked something out rather than asking for the result of their thinking. So, faced with a student who has written 28 + 53, a probing question would be, "How will you work out the answer?" as opposed to, "What is the answer?" Or, similarly, if a student has written 56–19 = 43, a probing question would be, "Tell me how you worked this out?" rather than asking the student to "Check this one again," which signals an error. By asking probing questions you come to understand the student's thinking processes rather than simply knowing whether they got an answer right or wrong.

When observing teachers at work, one way in which good subject knowledge is apparent is in the way that they deal with student's unexpected questions. Teachers with good subject knowledge are rarely thrown by these questions. They deal with them with confidence.

Finally, a teacher with good subject knowledge can support students in making the connections between different areas of mathematics so that the students see mathematics as a whole rather than a series of separate and often disparate ideas. They can do this because they themselves see mathematics as a whole. This is another example of Richard Skemp's idea of relational understanding.

How will this book develop my subject knowledge?

The three facets of subject knowledge are developed throughout the book. First, your knowledge of the curriculum is developed by beginning each chapter with a clear progression in terms of the mathematical concepts covered.

Each chapter develops your knowledge of the students and learning processes so that lessons and activities can be structured effectively to meet the needs of all students. This is done through the use of case studies and direct reference to how children learn mathematics. A focus on common misconceptions within each chapter also allows you to develop your understanding of how individual students come to understand mathematics.

The book also focuses directly on developing your knowledge of teaching and learning resources and how they can be used effectively to support learning. There are many ideas for tasks that you can use in the classroom. These both allow you to see how mathematics can be taught and give you ideas to try out for yourself. They mirror the process you will often go through to plan a session. You may be given a resource with which to teach a particular concept. As a teacher, your role is to plan the most appropriate way to use this resource to develop the children's mathematical understanding. Our aim is that this process will support you in becoming skilled designers of your own resources that you can tailor for the students that you teach. We think that by using the resources in your own teaching you will both develop your teaching repertoire and develop your subject knowledge. We all learn most about learning and teaching mathematics when we teach mathematics.

No book on subject knowledge could teach you everything you need to know across the whole of the curriculum. However, we hope that this book will ensure that you feel confident in all the key areas of mathematics. Perhaps more importantly, we hope it will give you the confidence to develop your own subject knowledge further in a range of curriculum areas through extending and generalising the ideas in the book with a focus on applying them in classroom situations. It will also allow and encourage you to reflect on your own thinking processes in order to develop your own subject knowledge.

A framework for teaching mathematics

You will see that within this book specific attention is paid to progression within the strands of mathematics. This is to allow you to plan over the longer term as you gain an understanding of what children's previous experiences will have been and how they will draw on your teaching to support them in the future. Rather than simply reproduce the strands from the current Australian Mathematics Curriculum, we have split the curriculum into eight broad areas that subsume the current strands. We hope this "future proofs" the book. Curricula change but mathematics does not. The broad areas are:

1   **problem solving using mathematics** This comes first as we think the most important facet of your subject knowledge is your ability to solve problems applying the mathematical understandings that you have

2   **number: counting; place value; fractions, decimals and percentages** This includes place value, fractions, decimals, percentages and ratio and proportion

3   **number: knowing and using number facts** This chapter will help you understand the underlying rules behind number patterns and number facts, including multiplication facts

4   **number: calculation (addition, subtraction, multiplication and division)** This includes number operations, mental methods for calculation and solving number problems

5   **algebra** This chapter covers spotting and creating patterns, understanding number patterns, finding equivalent relationships and solving simple equations such as missing number problems

6   **geometry** This chapter covers the properties of shapes such as **symmetry** and the language of position, including using **coordinates**

7   **measurement** This chapter covers conservation and comparison of measures, units of measurement, **area** and perimeter, and volume and capacity

8   **Statistics and probability** This chapter covers the data handling cycle including collecting data, analysing data and presenting data. It also explores chance and **probability**.

All the mathematical ideas that you are likely to meet when teaching the 4–13 age range appear within these chapters. The next chapter offers you a range of audit tools that allow you to assess your current subject knowledge against these broad areas. You can then prioritise your own learning and decide which chapters of the book to focus on first. At the beginning of each chapter the big ideas underpinning the mathematics within the chapter are detailed. This allows you to understand what the most important ideas are within a concept. It is amazing how often teachers have not been given an understanding of the big picture. As a result, their own subject knowledge consists of a rather disjointed series of tricks and rules. Having an understanding of this big picture allows you to understand the connections between areas of mathematics and develop your relational understanding.

Foundations for learning mathematics

Although the focus for this book is primary mathematics, it is important that all Early Years Educators feel secure in their knowledge of mathematics. The list below offers all Early Years Educators broad areas on which to focus to ensure that young learners begin to enjoy and become confident in their mathematics. Chapter 11 has been written to include the most recent developments in Early Years practice in Australia and explore teaching in the Early Years in more detail. In particular, we emphasise that mathematics learning should take place through child-initiated activity and discussion. The task of a good teacher of mathematics in the Early Years is to plan a wide range of activities that offer young learners opportunities across the breadth of the mathematics curriculum. As an Early Years Educator your assessments will allow you to track children's learning in the different areas of mathematics. Then your mathematical subject knowledge will allow you to make useful and pertinent interventions to support your learners' mathematical development.

Focusing on the following mathematical objectives will allow your learners to develop a broad understanding of mathematics. This list corresponds to the Foundation Year achievement standards and will be discussed further in Chapter 11:

- counting reliably with numbers up to 20
- connecting number names, numerals and quantities
- being able to subitise small collections of objects
- comparing numbers in collections of objects (up to 20)
- using practical materials to model addition and subtraction
- using everyday language and direct and indirect comparison to talk about size, weight, capacity, position, distance and time
- connecting days of the week to familiar events
- comparing and classify objects by their properties
- recognising, creating and describing patterns
- exploring characteristics of everyday 2D and 3D objects and shapes and using mathematical language to describe them
- Answering Yes/No questions to collect information.

Our aim is that by using this book Early Years' Educators will develop the knowledge and understanding to teach these concepts well. It is not the case that Early Years Educators need a less well-developed mathematical subject knowledge than their colleagues who teach older children. On the contrary it is vital that Early Years Educators have solid subject knowledge across the mathematics curriculum so that they can see how these early mathematical concepts develop. In order to plan for mathematics to be developed in a broad range of contexts and to plan for children to gain confidence and competence across the very broad range of standards, teachers need to have a secure subject knowledge themselves.

Organisation of the book

The second chapter of the book allows you to audit your subject knowledge in three ways. You can use these initial audits to create an action plan that will personalise your study to develop your subject knowledge in a way that suits your own individual needs.

Chapters 3 to 10 are structured in the same way so that you can easily manoeuvre your way around the book and so that you can see your subject knowledge developing across the whole mathematics curriculum. Each chapter is introduced using a starting point. Here you will be asked to engage in an activity that illustrates the main mathematical ideas underpinning the theme of the chapter. This is followed by a detailed breakdown of the progression within the mathematical ideas that are covered in the chapter. This supports the development of your curricular subject knowledge and allows you to plan appropriately for all of your learners, whatever age or stage they have reached.

The subject knowledge that underpins the strand is then discussed in more detail, explaining the **Big Ideas**, with particular care taken to highlight key terms, which are highlighted in the text when they first appear. These key terms are gathered together as a glossary at the end of the book. Careful explanations of the key areas of mathematics are also given so that you are able to understand how to explain important areas of mathematics to students. The bulk of each chapter is constructed around key teaching points. These teaching points often use a common misconception as a starting point for the discussion of a mathematical idea together with suggestions of how best you may introduce this idea to your students. The chapters also offer examples of activities that you can use in your classroom. This allows you to develop your own bank of tasks whilst developing an understanding of how best to use these activities. If you buy the companion volume you will find these tasks developed into full lesson ideas.

Each chapter also includes an **In practice** section. Here you are given an exemplar lesson plan focusing on a key concept from the mathematics covered in the chapter. The In practice section also includes a commentary by the teacher on the effectiveness of the lesson when they taught it.

Each chapter contains **Portfolio tasks** which can be used to help you reflect on your learning and can be included in any portfolio you need for your assessment on a teacher training or professional development course. In addition, each chapter contains sections that direct you to the most important research exploring the learning and teaching of this particular area of mathematics. These "Taking it further" sections are contained in boxes labelled **From the research** for articles that take a research focus and **From the classroom** for articles written with a classroom focus. These features will be of specific interest to those of you who are studying at Masters or a more advanced level.

At the end of each chapter, you are offered ideas for assessment in the particular area you have been studying. This section consists of a range of tasks and activities you can use in your classroom to assess your learners' understanding of the concepts you have been exploring. There are also ideas for "cross-curricular projects," which will allow you to explore the ideas introduced in the chapter in an extended project that draws on other areas of the curriculum.

Finally, there is a **Self-audit** and you are also invited to construct a lesson plan focusing on one facet of the subject knowledge covered in the chapter. Your response to these two activities could form part of a portfolio.

Chapters 11–13 further develop your expertise in teaching mathematics, your pedagogical subject knowledge, through exploring subject knowledge in terms of Early Years practice, inclusive teaching and learning, the use of calculators and using ICT to support the learning and teaching of mathematics. All include the most recent research and developments.

The companion website

Everyone who purchases the book is given access to the companion website. It contains the initial audit activities and the proforma for your personal action plan, as well as templates for lesson plans. You will also find a summary presentation of each chapter that can be used by teacher trainers using the book to support their teaching. Some chapters are also supported by a video clip of learners working on the mathematical activities in the classroom or short interviews with classroom teachers discussing how they have used these activities in their classrooms. In addition to this, academic papers that are discussed in the book are available on the site together with additional suggested reading.

Summary

This chapter has twin aims: to describe to you what is meant by subject knowledge and to outline to you how the book will support you in developing your mathematical subject knowledge in order to become a confident teacher of mathematics. You have seen how subject knowledge can be described as having knowledge in three areas:

- mathematical subject matter knowledge (you understand the mathematics)
- curriculum subject knowledge (you know the requirements of the curriculum)
- pedagogical subject knowledge (you can make good choices to help you teach the mathematics).

And you have been introduced to the key features in the book that will allow you to develop across these areas. Your next task is to complete the audit in the next chapter and then enjoy yourself as you draw on the appropriate chapters to complete your personal portfolio.

Going further

Diamonds in a skull: unpacking pedagogy with beginning teachers by Tony Cotton

This research paper is based on the opening of this chapter. It describes a piece of research that involved 20 pre-service teachers in describing and exploring how they developed their subject knowledge in mathematics. These pre-service teachers had described themselves as lacking confidence in teaching mathematics, and the research tracks their development in becoming skilled teachers of mathematics.

Cotton, T. (2010) 'Diamonds in a skull: Unpacking pedagogy with beginning teachers', in Walshaw, M. (ed.) *Unpacking Pedagogy: New Perspectives in Mathematics Classrooms*. Charlotte, NC: Information Age Publishing.

What teachers need to know to teach mathematics: an argument for a reconceptualised model by Derek Hurrell

This paper looks at alternative models of teacher subject knowledge in the Australian context. There is a good summary of Shulman's model of subject knowledge as well as other models. The paper offers a revised mathematical knowledge for teaching model.

Hurrell, D. (2013) 'What teachers need to know to teach mathematics: An argument for a reconceptualised model', *Australian Journal of Teacher Education*, 38(11).

Relational understanding and instrumental understanding by Richard Skemp

This book is seen as one of the most important pieces of writing about mathematics teaching and learning of the last 50 years. Skemp uses a metaphor that describes mathematics as a landscape to be explored rather than seeing mathematics as a ladder to be climbed. Through gaining an understanding of how different points in a landscape are connected and the different paths we may take to navigate this landscape, we come to a much better understanding than if we have to blindly follow paths without any sense of where they might lead. The full text of the article is available at http://math.coe.uga.edu/olive/EMAT3500f08/instrumental-relational.pdf and is well worth reading.

Skemp, R. (1976) 'Relational understanding and instrumental understanding', *Mathematics Teaching*, 77(3), 20–26.

Teacher knowledge as fundamental to effective teaching practice. A special edition of the Journal of Mathematics Teacher Education edited by Mary Walshaw

This special edition of the journal reports on research from England, Turkey, Australia and the United States to argue that teacher subject knowledge is central to the development of teacher effectiveness. The articles show that teachers' choices around teaching strategies rely on their personal knowledge of the content to be taught. The full journal can be downloaded from https://link.springer.com/content/pdf.

CHAPTER 2

WHAT SHOULD I KNOW?
WHAT DO I KNOW?

Introduction

You have probably all experienced something like this. At the end of an explanation, which in your mind was very clear, one of the children you are working with will put their hand up. When you ask them what they want they will tell you "I'm stuck." You might reply, "What are you stuck on?" "Just everything," they say.

Developing your subject knowledge in mathematics can feel a bit like this; it is hard to focus on the particulars when at first you are not sure what you understand and what you need to work on. Tony remembers his father as his best teacher. He remembers him being able to explain things really clearly, especially the mathematics he was struggling with. He remembers that it was only recently his father told him that he did not always understand the mathematics himself, but that he just encouraged Tony to explain things to him until he was able to discover for himself where he was stuck.

If you find yourself struggling on any of the questions in this chapter, try to see this as a positive. You have found an area that you need to work on. Carefully reflect on what you do understand and ask yourself exactly what questions you have. If you can identify your questions carefully you may well be able to move forwards yourself. Indeed, accept "stuckness" as a natural part of learning. In some sense, we cannot learn anything without being stuck. If we never have to stop and think, if we are not challenged, then we are simply completing exercises that we already know how to do. And what is the point of that?

Chapter 1 described what we mean by subject knowledge. This chapter expands on this by offering you audit tasks so that by the end of the chapter you have a clearer understanding of the mathematical knowledge you already have and the steps you might take to build on your previous experiences.

DOI: 10.4324/9781003315155-2

Starting point

Two groups of pre-service teachers were discussing the ways that their courses assessed their subject knowledge to ensure that they could teach mathematics effectively. Neither group were happy. One institution asked its pre-service teachers to complete a formal assessment at the end of each year. This group asked Tony,

> *How does a test show that we can teach mathematics well? If we passed our mathematics examinations at the end of school, why should we have to do more tests?*

The other group were equally unhappy. They had to complete subject knowledge files and felt that gathering evidence to show that they had revised all the mathematics that comprised the school curriculum did not support them in their teaching either. They felt that although this file contained evidence that they had covered all the areas of mathematics it did not necessarily show that they could use this knowledge to teach others. They argued that being able to answer a question in an assessment is not the same as having the deep understanding of a mathematical concept that allows us to teach it well.

It is clearly important that pre-service teachers are confident in their mathematical subject knowledge before embarking on teaching particular mathematical ideas. However, what the previous point shows is that it is also important that you can see and make explicit the connections between your own subject knowledge and your teaching.

During an observation of one of their placements, the supervising teacher asked these two questions of the pre-service teachers:

- how do you know what mathematics you should know?
- how do you know if you know it?

These are helpful questions in terms of assessing our own subject knowledge. So, answering these questions and making connections between your answers and your teaching is our aim for this chapter.

Auditing your current knowledge

This chapter will ask you to consider your subject knowledge in three ways. First, you will reflect on your previous experiences of learning mathematics and think about your current levels of confidence in teaching mathematics. It is important to explore this area in an initial audit as confidence is a key facet of good subject knowledge. The audit will allow you to recognise your developing confidence as one measure of your improving subject knowledge.

The second part of the audit explores your personal beliefs about what makes up effective mathematics teaching. Your own beliefs play a central role in the way in which you teach mathematics, and this again links back to how confident you feel when teaching mathematics. As we have already suggested, feeling confident in the classroom is a large part of demonstrating good subject knowledge. This section of the audit allows you to illustrate how you will develop in order to plan appropriate activities in order to ensure effective learning of mathematics. Finally, you will audit your current understandings of mathematics in the areas of:

- problem solving using mathematics
- number: counting; place value; fractions, decimals and percentages
- number: knowing and using number facts
- number: calculation (addition, subtraction, multiplication and division)
- algebra
- geometry: properties of shapes; position and direction
- measurement
- statistics and probability.

These areas correspond to the chapters that follow in the book. This section of the audit takes the form of presenting you with common misconceptions for you to analyse in the areas listed earlier. Using the proformas available on the companion website you will be able to complete these audits online and they will form the introductory section of your personal e-portfolio of mathematical subject knowledge.

Audit: section 1 – previous experience in learning mathematics and confidence in teaching mathematics

This section of the audit documents your previous experience in learning mathematics and allows you to document your developing confidence in teaching mathematics and your planning for teaching. It will also give you areas which you need to research further to develop your own subject knowledge.

Visit the companion website to complete the audit online. You will see that the audit is in two columns. Complete the first column before using the book to work on your subject knowledge and then return to this section and complete the second column after using the book. This allows you to document any changes in the way you view your previous experience of learning mathematics as well as any changes in your confidence levels as a result of your studies.

The comments do not have to be detailed. For example, you may not have many memories of your primary school in order to make comments. The aim of this section is to allow you to take some time to think through your prior experiences and achievements and to reflect on how this makes you feel currently, as a learner and teacher of mathematics. The second part of this opening section allows you to record statements about how you feel about teaching, planning and working on mathematics. The aim here is that you can take note of how your feelings develop as a result of your studies.

The final part of this section explores your initial confidence levels across the mathematics curriculum and then allows you to reflect back to these initial thoughts after you have worked through the activities in the book. If you are uncertain as to the content of any of these areas of mathematics turn to the appropriate chapter and skim through the **progression** section. This will remind you of the mathematical content contained within each of these areas.

An example of a completed audit is shown later. If you wish, print off this initial audit and place it in a file as the opening piece of evidence of your developing mathematical knowledge.

Mathematics Subject Knowledge – Personal Profile		
Name	Michael Minas	
Previous experience and qualifications	Comment on your learning and achievement at the following stages	
Age 11:	Felt very confident in learning maths. Always found it enjoyable. I had a teacher who made strong links between the real world and what we were learning about in class, thus the subject seemed important and meaningful to me.	I look back now and realise how important these real-world connections were, as they helped me to see mathematics as a practical subject that allowed me to make sense of the world around me.
Age 15:	Moved to a new high school and for the first time in my life, maths didn't come easily to me, relative to my peers. I remember not being able to follow what we were doing in a Year 10 maths class and believing that I was one of the few students in the class who were confused. I found this to be quite confronting, as I was so used to completing tasks with ease.	I look back now and realise a couple of things. First, I have no idea whether I was the only student in the class that was struggling, that was just my perception at the time. Perhaps if I had been more aware of other students who were finding the work challenging, I may have been more inclined to apply myself more, rather than become discouraged.

Mathematics Subject Knowledge – Personal Profile		
Name	Michael Minas	
		I also think the fact that the first time I ever felt really challenged came so late in my schooling journey was probably not a good thing. If I had spent more time feeling truly confused earlier in primary school, perhaps I would have been more comfortable working my way to clarity.
Year 12	I did fairly well in Year 12 maths, but I was continuing to lose my sense of enjoyment and wonder, as I saw very few practical applications.	I realise that I feel a bit frustrated that some of my secondary teachers did not make the subject enjoyable or interesting.
Post Year 12	I have maintained a growing interest in maths ever since I finished high school, which has grown ever since I started to work with primary students.	
My confidence in teaching mathematics	Complete the following sentences Before using *Understanding and Teaching Primary Mathematics*	After using *Understanding and Teaching Primary Mathematics*
When teaching mathematics in school I feel . . .	OK. I do seem to just follow the school's plans or use ideas from the web rather than thinking up ideas for myself.	That it is important for me to have a really good grasp of the maths rather than simply read through the maths at the same level as the children – it helps me see where the maths is leading in the future.
When planning for teaching mathematics I feel . . .	See previous	That I need to ensure the activities are challenging and interesting – not simply find something that will cover the curriculum
When researching mathematics in order to develop my own subject knowledge I feel . . .	That I can find activities that will help the students. I find it hard to find exciting activities, though	See previous

Audit of confidence in mathematical subject knowledge

The second audit available on the companion website supports you in thinking about your current confidence across all the areas of mathematics that are covered in the book. Think about each of the areas. Again, if you are not sure what mathematical ideas are contained in an area flick to the chapter and look at the **progression** section. This outlines the content covered in a particular area in detail.

Please complete this audit using the following key:

1   I feel very comfortable with this area of mathematics and could answer questions from children with confidence
2   I feel fairly comfortable with this area of mathematics and would be happy to teach it with some preparation
3   I am a little uncertain with this area of mathematics and would need to spend a lot of time preparing before I could teach it
4   I do not understand this area of mathematics at all.

An example of a completed audit is given on the following page.

Area of mathematics	Before reading book		After reading book	
	Confidence level	Comment or evidence	Confidence level	Comment or evidence
Using mathematics	3	I'm not really sure what is meant by "using and applying" as it is an area I feel I did not learn in school.	2	I feel much more confident that I understand what I need to teach – I still find it a challenge to find appropriate activities, though. I have included my response to the initial audit task here as well as my notes from reading the chapter to show how my understanding has developed.

Area of mathematics	Before reading book		After reading book	
	Confidence level	Comment or evidence	Confidence level	Comment or evidence
Counting; place value; fractions, decimals and percentage; ratio and proportion.	2	This is an area I feel OK about – it has been an area I have previously focused on. My successful response to the initial audit tasks supports me in this comment.	2	I did not focus on this area.
Knowing and using number facts	2	As previously	1	I read the chapter and found the In practice section interesting. I have included a lesson plan in my portfolio that I have used in school. I am now very confident in teaching this area of mathematics.
Calculating	2	As previously	1	I also completed the tasks at the end of the chapter and was very pleased with the results. I have included these as evidence of my growing confidence in this area.
Area of mathematics	Confidence level	Comment or evidence	Confidence level	Comment or evidence
Algebra	3	I only remember doing algebra at secondary school so was surprised to see it included. I thought it would be too difficult for the children I teach.	2	After reading the chapter and trying the tasks I realise that I did not really realise that algebra starts very early. In fact, a lot of what I already taught was algebra. I feel much better about this area now.

Area of mathematics	Before reading book		After reading book	
	Confidence level	Comment or evidence	Confidence level	Comment or evidence
Geometry	2	This is another area I feel comfortable with. I have a good memory for the names and properties of shapes. I have included my successful initial audit as evidence of this.	2	After working on the chapter, I realise that there is a lot more to shape than simply remembering the names. I really enjoyed working through this chapter and have included all my notes in my portfolio.
Measurement	3	I am uncertain about areas and volumes and need to study this in some detail.	2	I feel much more confident with my knowledge in this area. I have included my response to the tasks in the chapter and an exemplar lesson plan in my portfolio.
Statistics and handling data	3	This is not an area that I feel confident in at all as it is something that I avoided at school and have not had to teach at all.	2	After working on the chapter, I realise that this area is not as complex as I thought. I have included my notes from working through the chapter as evidence of my improvement.

Audit: section 2 – beliefs about learning and teaching mathematics

For this section of the audit you are asked to reflect on your beliefs about learning and teaching mathematics and comment on any changes to your thinking after working through the book. You do this through completing a questionnaire based on important research into effective teaching of mathematics carried out by Mike Askew and others at King's College in London, England (for full details see "Going further" at the end of this chapter). They carried out a wide range of observations across many primary schools exploring the impact of teachers' beliefs on learning mathematics and divided these beliefs into three categories: "connectionist", "transmission" and "discovery." To summarise these beliefs:

Connectionist teachers think it is important to make connections between all the different areas of mathematics. They are likely to draw on a wide range of mathematical ideas to solve problems themselves and will make these links for their learners. These teachers will favour open-ended, investigative approaches to learning and teaching mathematics.

Transmission teachers take the view that mathematics is a fixed body of knowledge that their learners should be introduced to by a teacher who can pass on this knowledge effectively. A model here might be that the teacher works through examples with their learners who then practise similar examples on their own.

Discovery teachers will plan activities that allow their learners to explore mathematics at their own level, coming to their own understanding of how mathematics operates.

Your own beliefs are likely to have been influenced by your prior experiences of learning mathematics and how this makes you feel as a learner of mathematics. They will also have been influenced by your experiences in the classroom as a trainee teacher and the beliefs of teachers you have worked with in schools.

Once you have completed the initial questionnaire you may like to explore the differences between your own beliefs and those that you would ascribe to your own teachers and to teachers that you have worked with in schools. Differences in beliefs may account for personal feelings of inadequacy as both a teacher and a learner.

Visit the companion website containing the questionnaire and highlight the statements that most closely mirror your current beliefs using the highlighter tool. You can select statements from different columns. In fact, it is very likely that you will want to select statements from across the spectrum of beliefs. You may even want to choose more than one statement per row. The idea is to engage you in thinking about your beliefs before you embark on working with the ideas in this book. When you feel as though you have completed working through the book revisit the questionnaire and underline those statements you now agree with. Of course, many of these may be the same. Then complete the comment box to reflect on the impact of your working at your mathematical subject knowledge on your beliefs about learning and teaching mathematics.

Look at the completed table as an example before completing the questionnaire.

	Connectionist	Transmission	Discovery
Beliefs about what it is to be a numerate student	Being numerate involves:	Being numerate involves:	Being numerate involves:
	The use of methods of calculation which are both efficient and effective	Primarily the ability to perform standard procedures or routines	Finding the answer to a calculation by any method
	Confidence and ability in mental methods	A heavy reliance on paper and pencil methods	A heavy reliance on practical methods
	Selecting a method of calculation on the basis of both the operation and the numbers involved	Selecting a method of calculation primarily on the basis of the operation involved	Selecting a method of calculation primarily on the basis of the operation involved
	Awareness of the links between different aspects of the mathematics curriculum	Confidence in separate aspects of the mathematics curriculum	Confidence in separate aspects of the mathematics curriculum
	Reasoning, justifying and eventually proving results about number	An ability to "decode" contextual problems to identify the particular routine or technique required	Being able to use and apply mathematics using practical apparatus
Beliefs about students and how they learn to become numerate	Students become numerate through purposeful interpersonal activity based on interactions with others	Students become numerate through individual activity based on following instructions	Students become numerate through individual activity based on actions on objects
	Students learn through being challenged and struggling to overcome difficulties	Students learn through being introduced to one mathematical routine at a time and remembering it	Students need to be ready before they can learn certain mathematical ideas
	Most students are able to become numerate	Students vary in their ability to become numerate	Students vary in the rate at which their numeracy develops

	Connectionist	Transmission	Discovery
	Students have strategies for calculating but the teacher has responsibility for helping them to refine their methods	Students' strategies for calculating are of little importance – they need to be taught standard procedures	Students' own strategies are the most important: understanding is based on working things out yourself
	Students' misunderstandings need to be recognised, made explicit and worked on	Students' misunderstandings are the result of a failure to "grasp" what was being taught and need to be remedied by further reinforcement of the "correct" method	Students' misunderstandings are the results of students not being ready to learn the ideas
Beliefs about how best to teach students to become numerate	Teaching and learning are complementary	Teaching is separate from and has priority over learning	Learning is separate from and has priority over teaching
	Numeracy teaching is based on dialogue between teacher and students to explore understandings	Numeracy teaching is based on verbal explanations so that students understand teachers' methods	Numeracy teaching is based on practical activities so that students discover methods for themselves
	Learning about mathematical concepts and the ability to apply these concepts are learnt alongside each other	Learning about mathematical concepts precedes the ability to apply these concepts	Learning about mathematical concepts precedes the ability to apply these concepts
	The connections between mathematical ideas need to be acknowledged in teaching	Mathematical ideas need to be introduced in discrete packages	Mathematical ideas need to be introduced in discrete packages
	Application is best approached through challenges that need to be reasoned about	Application is best approached through word problems that offer contexts for calculating routines	Application is best approached through using practical equipment

	Connectionist	Transmission	Discovery
Comment on impact of self-study on personal beliefs	I think I have moved more towards the connectionist viewpoint. This was quite a difficult move for me as I have never been taught by a connectionist teacher and most of the teachers I have observed have focused on transmission. However, I am now much more able to see the links between different areas of mathematics and think it is important for students to be able to make this link too		
	I have also realised that using children's misconceptions as a starting point for teaching is really useful. I think my own growing understanding of where misconceptions come from has helped me		

Audit: section 3 – exploring subject knowledge

The final section of the audit asks you to work with a range of questions across the strands of mathematical subject knowledge covered in the books. This allows you to decide which areas of mathematics you already feel comfortable with and those areas you need to concentrate on for further study. Again, visit the companion website to carry out this section of the audit. The questions and worked answers are available on the companion website.

Do not worry at all if you find yourself thinking, "I don't know what to do in this question" or if you become stuck. This is a signal for you that it will be worth working through the appropriate chapter in the book. This is an initial audit of your subject knowledge so that you can focus on the areas you most need to develop. Once you have worked through all the questions you can complete a personal action plan on the companion website, which will help you to prioritise your use of the book. You should make your first priority those areas that you have assessed as 3 or 4.

Summary

This chapter has asked you to reflect on your own starting point in terms of your subject knowledge of mathematics. We hope that you are now able to answer the questions for yourself:

- what do I need to know?
- what do I already know?
- which areas should I focus on to develop my own knowledge?

The audits have provided you with the opening sections for your personal portfolio. You will develop this portfolio as you complete the portfolio tasks throughout the chapters you focus on and the audit sections at the end of these chapters.

Going further

Effective teachers of numeracy in primary schools: Teachers' beliefs, practices and students' learning by Mike Askew, Margaret Brown and colleagues

Mike Askew, Margaret Brown and a team of researchers from King's College observed and interviewed a large number of teachers to try to find out what beliefs correlated with effective teaching of mathematics. They explored teachers' beliefs about the nature of mathematics; their beliefs about the way that mathematics should be taught and finally their beliefs about how mathematical achievement should be assessed. The final report makes fascinating reading.

Askew, M., Brown, M., Rhodes, V., Wiliam, D. and Johnson, D. (1997) *Effective Teachers of Numeracy in Primary Schools: Teachers' Beliefs, Practices and Students' Learning*, paper presented at the British Educational Research Association annual conference in 1997, King's College, London.

Principles of practice and teacher actions: influences on effective teaching of numeracy by Tracy Muir, The University of Tasmania

Tracy Muir from the University of Tasmania discusses three case studies of teacher of mathematics in Australia and develops the "transmission, discovery, connection" to identify a set of principles and actions which can provide a framework for describing effective teaching of mathematics.

Muir, T. (2008) 'Principles of practice and teacher actions: Influences on effective teaching of numeracy', *Mathematics Education Journal*, 20(3), 78–101.

CHAPTER 3

PROBLEM SOLVING USING MATHEMATICS

Introduction

What follows is a true story, although whenever we share it with pre-service teachers, they tell us that they do not believe it really happened. However, if we were to suggest that if we sit reading a book to ourselves in the reading corner and that very quickly my class would start to gather round to share the book they can see that this might have happened. We think children can find mathematics as fascinating as reading. It is important that we show the learners we work with that we enjoy doing mathematics for its own sake. The story that follows suggests one way of doing this.

The class teacher had been experiencing difficulty in engaging a group of nine- and ten-year-olds in a problem-solving activity. They seemed happy simply completing exercises they had been given but would not investigate mathematics in a more open way. So, Tony, who was working with the teacher, decided on a new tactic. Whilst the group were out at break, he surrounded himself with large sheets of flip chart paper and began to explore a particular number chain. (The rule was that if the number is divisible by 5 it should be divided by 5 otherwise you add 4. The aim is to see how long the chains were depending on the starting number.) As the group came back into the classroom, they started to gather round and asked Tony what he was doing. They then started to ask other questions, "How long do you think this chain will be?" "What happens if you start with a number bigger than 10?"

They made further suggestions about how they could help in finding the answers to their questions. Before long, the whole class was engaged in exploring number chains. Sometimes exploring mathematics alongside our students engages more effectively than always working at something that we already know the answer to.

The aim of this chapter is to support you in becoming confident in teaching how to problem solve using mathematics. To do this

DOI: 10.4324/9781003315155-3

you need to explore your own learning of mathematics. We would like to ask you to work on a piece of mathematics as an initial step. This will give you some insight into how your learners may feel when you ask them to carry out a mathematical activity they have not met before. By reflecting on your learning process, you will be better able to see the support you might offer to your students.

Portfolio task 3.1

Look at this number square.

Imagine you can extend it as far as you like both horizontally and vertically.

You may want to sketch your own enlarged version in a note-book.

1	3	5	7	. . .
2	6	10	14	. . .
4	12	20	28	. . .
8	24	40	56	. . .
. . .	. . .	. . .	. . .	. . .

What patterns do you see in the number square?

Jot down anything you notice.

Spend some time on this before you read on.

Now let us focus your thinking a little.

- do you think every number will appear if you extend the square far enough?
- if you think every number will appear, write down why you think this is the case
- if you don't know what to write, talk to somebody about it and then hone your ideas so that you can write something down
- will every number appear once, or more than once? Again, try to make notes describing your reasoning.

Now, to focus even more, do you think 1,000 will appear in an extended version of the square? If you think it will, describe its position. If you think it won't, explain why this is the case.

If you use this activity with students some may tell you that they find the first question too open. They may say, "I don't notice anything" or even, "What am I supposed to notice?" Eventually a group of students may make a suggestion like: "If you look at a block of four numbers from the grid, by adding the two numbers in the first column you get the number at the top of the next column."

1	3
2	6

Once someone has noticed a pattern, any pattern, this seems to give other students permission to look for other patterns. Perhaps at this stage students realise they are not being asked to find a particular pattern that will give an answer that is in the teacher's head, but that they are genuinely asked to describe any patterns they can see.

Unlike the other chapters in this book, the ideas that underpin problem solving cannot be summarised as a set of content to be taught; rather it is a structure that you can use to solve and investigate problems across all areas of mathematics. Later in the chapter, you will see that ideas of communication, reasoning and enquiry are at the heart of this mathematical problem solving. It is for this reason that we have chosen to open the content knowledge section of the book with this chapter. We think these ideas are at the heart of all mathematical activity. If you become confident and skilled in the areas of mathematical communication; if you are able to use your mathematical knowledge to investigate and solve problems through applying mathematical reasoning, you will have become a mathematician.

The Australian Curriculum for mathematics contains four proficiency strands: understanding, fluency, problem solving and reasoning. The stated aim for the inclusion of these four elements is to ensure that the mathematics presented to students is relevant to the 21st century. In this sense the proficiencies are seen as 21st-century skills. The curriculum document suggests that these proficiency strands will:

> *enable students to respond to familiar and unfamiliar situations by employing mathematical strategies to make informed decisions and solve problems efficiently.*
> (Australian curriculum: Proficiency strands)

To expand on these strands "understanding" refers to being able to make connections between related ideas in mathematics and to draw on these connections as we explain our mathematical thinking to others. This is what connectionist teachers (see Chapter 2) emphasise. "Fluency" is described as the ability to choose appropriate procedures and carry them out accurately when we work on mathematical problems. Both fluency and understanding support is when we engage in "problem solving." That is, we can look at and represent a problem in such a way that we can see what mathematical skills we can employ in order to solve the problem. We can apply these methods efficiently and explain what we are doing clearly to someone else. We can look at this another way. Through engaging in problem solving activities, we can develop our understanding and fluency. The fourth proficiency is "reasoning." This is described as "analysing, proving, evaluating, explaining, inferring, justifying and generalising." As you read further in this chapter you will see that we see these ideas as at the heart of problem solving.

So, two of the proficiencies directly refer to problem solving and reasoning, and the proficiencies are described as "an integral part of the mathematics content" across all of the content strands. So, it seems appropriate that problem solving should be our starting point.

If you tried to share your insights into the previous number square task with friends or students you will have started to see the centrality of communication. Such communication is embedded in the mathematical proficiencies described previously. Sometimes, our own thoughts and ideas become much clearer when we try to share them with others, or we gain new insights through asking questions as other people try to explain their thinking. You may have realised that the prompt questions asked you why you thought your solutions were correct. This is to enable you to begin to articulate your reasoning. It is only when we ask "why" that we begin to explore mathematical reasoning. Finally, the process of enquiry and problem solving encourages learners to engage with open questions. You were asked, "What patterns can you see?" This may have felt too open for you at first. How would you know when you had finished? Were there particular patterns you were expected to spot? This unease perhaps comes from your prior experience of mathematics, which may have been presented through a series of closed questions and answers supplied by the teacher, followed up by a series of exercises to practise a particular skill. When developing problem solving skills, we often start with a fairly open question so that we can investigate the mathematics in a number of different ways.

Before moving on to explore problem solving in more detail, we will outline how the previous ideas can be developed throughout a student's time in primary school. This allows you to see how the activities you design for the children you are working with build on ideas that they will have met already. Perhaps you can see how the task above began to model the process of supporting students to develop their problem solving skills. It showed how you can break problems down into simpler steps. It was also encouraging you to persevere. In fact, it was expecting you to become resilient, to not jump to the first answer you thought of but to explore different routes through the problem.

The International Baccalaureate Programme for Primary Schools expects learners to be:

- exploring, wondering and questioning
- taking and defending a position
- using critical thinking skills to understand a concept
- making and testing theories
- experimenting and playing with possibilities
- solving problems in a variety of ways.

Looking back at the portfolio task we hope that you can see your thinking developing along these lines.

From the research

In the book *Care in Mathematics Education: Alternative Educational Practices and Spaces*, Anne Watson devotes a chapter to exploring mathematics education in a range of Indigenous communities, including describing the approach to mathematics learning in a school she calls Lochville, located in a large Australian city. She describes how teachers in this school have shifted their approach towards problem solving in which the focus of the curriculum is on building concepts rather than memorising methods. They are taking a collaborative approach to the learning using resources and materials that are appropriate to the Indigenous students they are working with. The extended quotation below shows the impact this focus on mathematical proficiencies is having on these Indigenous students:

> *The mathematical work is at first divergent, but is closed down to mathematical convention through dialogue, identifying efficient methods that "a mathematician would use", provision of conventional language and symbolism, and relating to each students [learning] goals. They use a pedagogical framework: launch, explore, summarise, reflect, apply or analyse. Much of this mirrors traditional ways of learning.*

You can see the mathematical proficiencies outlined earlier in the chapter in this approach. You will also recognise the pedagogical framework that is described in the "In Practice" lessons that close each chapter.

The statements in the details of progression show how you can gradually develop the skills in the learners that you work with. It also shows you how your own skills should develop. The statements are taken from the Australian National Curriculum but are applicable generally.

Progression in problem solving using mathematics

Foundations for problem solving using mathematics

Initially, young learners should be asked to use materials to model authentic problems, including sorting objects or using familiar counting sequences to solve unfamiliar problems. They should begin to explore the reasonableness of an answer. They will be developing their reasoning skills by explaining comparisons of quantities, creating patterns and explaining processes for indirect comparison of length.

Beginning to problem solve using mathematics

At this stage problem-solving includes using materials to model authentic problems, giving and receiving directions to unfamiliar places, using familiar counting sequences to solve unfamiliar problems and discussing the reasonableness of the answer. It also includes formulating problems from authentic situations, making models and using number sentences that represent problem situations and matching transformations with their original shape.

Reasoning includes explaining direct and indirect comparisons of length using uniform informal units, justifying representations of data and explaining patterns that have been created. Students will also use known facts to work on unfamiliar calculations, compare and contrast different calculation strategies and create and interpret simple representations of data.

Becoming confident in problem solving using mathematics

By this stage students are coming to an understanding of the way in which mathematics can be used to communicate and explain ideas. Problem-solving, at this stage, includes formulating and modelling authentic situations involving planning methods of data collection and representation, making models of three-dimensional objects.

Students will also be using number properties to continue number patterns; comparing large numbers with each other; comparing time durations and using properties of numbers to continue patterns

Reasoning includes using generalising from number properties and results of calculations, comparing angles and creating and interpreting variations in the results of data collections and data displays. Students will also be generalising from number properties and results of calculations; deriving strategies for unfamiliar multiplication and division tasks; comparing angles; communicating information using graphical displays and evaluating the appropriateness of different displays.

Extending learning in problem solving using mathematics

Students operating at this level will be able to formulate and solve authentic problems using numbers and measurements and creating financial plans using their problem solving skills. They will also be interpreting secondary data displays and finding the size of unknown angles

Reasoning, at this stage, includes investigating and explaining mental and written strategies to perform calculations efficiently; continuing number patterns involving fractions and decimals; interpreting results of chance experiments; posing appropriate questions for data investigations and interpreting data sets. Students will also be able to explain the transformation of one shape into another and why the actual results of chance experiments may differ from expected results.

The next section of the chapter focuses on the big ideas of communication, reasoning, and enquiry and problem solving. These big ideas are the starting point for all the mathematical problem solving we will carry out with our students. You will explore ways in which you can become a confident mathematical thinker, and, in this way, you will be able to model mathematical thinking successfully with the students with whom you work.

Big ideas

One of the most influential texts exploring this area of our subject is *Thinking Mathematically* by John Mason with Leone Burton and Kaye Stacey. John Mason was the Professor of Mathematics Education at the Open University in England. He has an international reputation and works globally sharing his ideas. He has a particular interest in the ways in which we search for generality as a way of thinking and making sense of the world.

In *Thinking Mathematically* we are offered five key assumptions to support our teaching of mathematical thinking:

1 Everyone can think mathematically
2 Mathematical thinking can be improved by "practising reflection"
3 Mathematical thinking is provoked by contradiction, tension and surprise
4 Mathematical thinking is supported by an atmosphere of questioning, challenging and reflecting
5 Mathematical thinking helps in understanding oneself and the world.

You may realise that these assumptions underpin the approach taken to developing your subject knowledge in this book and that they offer a classroom ethos rather than a series of skills to be taught. However, there are skills which underpin the process of problem solving using mathematics which will be outlined in this chapter.

Two other ideas which are important when working with children on their mathematical thinking are **specialising** and **generalising**. You might remember that the word "generalising" appears regularly in the Australian curriculum. However, generalising cannot take place without some "specialising."

Specialising is the process you use when you look at specific examples in order to get started on a problem. So, for example, in the opening activity, if you suggested that the "numbers doubled as they move down the rows" you will have started by noticing that 2 is double 1 and that 4 is double 2. You may even have checked that this was always the case by extending the number square to look at further examples. You will have used this to come to a generalisation. This is when you make a statement about all numbers in the square. So generalising is making a statement that is true about a wide range of cases. An example might be, "if you add two odd numbers together you get an even number."

The ideas of specialising and generalising appear throughout mathematics. They also underpin the mathematical proficiencies. But specialising and generalising do not always come naturally. A common confusion you may see in your students is that they will take a special case as true for all cases. If you were to ask your students to draw you a triangle, we would bet that they would draw an equilateral triangle. They are offering you a special case, a triangle with all sides and angles the same size, rather than seeing the bigger picture. In general, any three-sided shape is a triangle.

TAKING IT FURTHER: **FROM THE RESEARCH**

Kaye Stacey from the University of Melbourne (you may remember she worked on *Thinking Mathematically* with John Mason) shared her research on *What is mathematical thinking and why is it important* in a paper written in 2006. She suggests that solving problems with mathematics requires a wide range of skills to be developed including:

- deep mathematical knowledge
- general reasoning abilities
- knowledge of heuristic strategies
- helpful beliefs and attitudes
- personal attributes such as confidence, persistence and organisation
- skills for communicating a solution.

She also suggests that there are four fundamental processes that appear whenever mathematical thinking is taking place:

- specialising – trying special cases, looking at examples
- generalising – looking for patterns and relationships
- conjecturing – predicting relationships and results
- convincing – finding and communicating reasons why something is true.

Next time you engage in some investigational work, ask yourself which of these processes you noticed Similarly, the next time you work with children on developing their mathematical thinking skills, use the first list to evaluate your own practice.

Enquiry and problem solving

It may be useful at this point to explore a mathematical problem to exemplify what we mean by enquiry and problem solving.

Reflect on how you approached the earlier problem. We would encourage you to develop a classroom ethos in which students are exploring, wondering and questioning whenever they are working mathematically and that they will be experimenting and playing with possibilities in their learning. This suggests that it is not sufficient to teach children how to solve problems but that we should teach them to want to solve problems. Is this how you felt when faced with the problem earlier? Did you want to extend the problem and find out why certain results were happening, or did you feel rather confused by the apparent openness of the problem? Many of our students when faced with this problem will ask us how they will know when they have "Done enough" or "Finished." We would suggest that this attitude to problem solving is learnt. Many of you will remember having to submit coursework as a part of your examinations and many of you will have been given very clear guidelines as to the approach you should take as well as the way you should present the problem in order to maximise your mark. However, this places a reliance on the teacher to give you answers and to keep you on the right track, rather than an inclination to solve problems through following your own lines of enquiry.

An important way to develop your own subject knowledge is to explore mathematical problems for yourself. If possible, do this with the students you teach, so you can model how problem solving can be open-ended. Show them that you explore, wonder and question when faced with a mathematical problem. If your approach is to experiment and play with possibilities, they will follow your model. We offered a view of how this can be effective as a teaching approach at the beginning of the chapter.

Portfolio task 3.2

Look at the number line below:

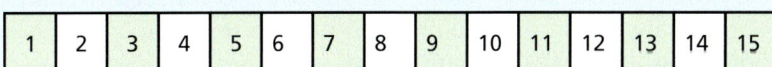

Look at chunks of the number line that consist of two consecutive numbers, such as 1, 2 or 6, 7 or 13, 14.

Add these pairs of numbers together.

- What do you notice?
- Do you think this result is true for any two consecutive numbers?
- Why do you think this happens?
- Now try adding together three consecutive numbers, or four. What are you noticing?
- What do you think is the reason for these results?

You may still feel a little insecure with the openness of this task. You may not be sure if you have answered the question. This is not unusual. There is in fact no question to answer. The point is to plan a route through the problem using your knowledge of similar problems. This is what Kaye Stacey means by knowledge of heuristic strategies. Hopefully, by trying to answer the "Why" questions you are beginning to "communicate your mathematical thinking." One of Kaye Stacey's fundamental processes.

The last task modelled a line of enquiry in the following way:

1 Start with a question – what do we notice about consecutive numbers?
2 Find a starting point to examine this question – in this case look at pairs of numbers
3 Specialise – that is, look at specific pairs of numbers to see what you notice. You may have noticed that adding together pairs of numbers gave you an odd number as an answer
4 Try a range of similar cases to see if the result can be generalised – does the result always happen whichever pairs you choose?
5 Try to "prove" the result – why does this happen? Proving is being able to convince someone that your explanation is correct.

Time for another task.

Portfolio task 3.3

Tony has used this problem with young children, and they proved the result for by building towers from cubes.

Try to prove the statement, "'The sum of two consecutive numbers is always an odd number' through a drawing." Here is a drawing that may get you started:

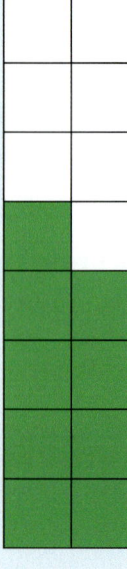

Representing

In the book *Assessing Children's Mathematical Knowledge: Social class, sex and problem solving*, Barry Cooper and Mairead Dunne showed that many students fail to demonstrate their skills in problem solving as they find it difficult to decode representations of mathematical problems. Students who are most successful in solving problems are those who can make the link between a realistic mathematical problem and the mathematical processes they will need to carry out to solve the problem.

The best way to support students in developing their skills in decoding different representations of problems, in words, in tables, in graphs, in pictures, is to allow them to explore a range of problems that are represented in a variety of ways. This range of representations can then be shared with the class so that your students begin to understand the range of representations that are possible. Whilst they are exploring the range of ways in which they may represent problems, they also need to be developing their skills in communicating their mathematical thinking, which is discussed later.

Reasoning and communication

Reasoning is the process of being able to plan an enquiry. That is to decide how you will tackle a problem, then drawing on the range of mathematical skills that you have available to work at the problem. You can develop reasoning by asking children questions that allow them to draw on the knowledge they already have to solve a problem and, most importantly, by encouraging them to articulate their thinking processes. Think back to the portfolio task that opened this section. Could you explain the thought processes you went through? In *Thinking Mathematically*, John Mason and his colleagues describe a process of rubric writing to help develop mathematical reasoning and communication. Rubric writing is following a routine every time you notice you are stuck. The key processes in rubric writing are noticing when you are stuck and writing down how you know you are stuck, what is it that you do not know? When you have a new idea, described as an "Aha" moment, you write down "Aha" and describe the discovery you have made. Finally, at regular intervals, you pause to check your calculation or reasoning and make note of this. Then, at the end of a process of problem solving you take time to "reflect" and notice what you have discovered and what you have learnt.

Portfolio task 3.4

Take a sheet of A3 paper and fold it in half. Use a portrait orientation. On the left-hand side of the paper revisit the consecutive numbers problem. Try to re-create in as much detail as you can the thinking process you went through when you explored the problem.

Now on the right-hand side of the paper use the method of rubric writing to annotate your thinking:

- When were you stuck?
- Why were you stuck?
- Jot down your "aha" moments.
- What did you do immediately following these thoughts?
- At what stages did you check back over your calculations?
- What did you discover?

Finally, take time to reflect on:

- what did you learn from working at the problem?
- how might you approach a similar problem in the future?
- how might you explore this problem further?

Try to develop your skills of "rubric writing" as you work on the tasks in the rest of the chapter.

If you can try this problem with a class you are teaching, encourage them to tell you when they are stuck or have "aha" moments. Support them in checking back over their solutions and most importantly give them time to reflect on the process.

Teaching points

When you introduce your class to mathematical problem-solving activities you will notice misconceptions and misunderstandings. It is important that you do not treat these as mistakes or errors but that you use them as prompts to intervene and to support the students in coming to a better understanding of the mathematics they are working on. The following teaching points are based on our experience of exploring problem solving tasks with students. You may recognise some of the areas as ones that you have difficulty with yourself. You will also see how they link to the big ideas described earlier.

Teaching point 1: not carrying out appropriate steps in multi-step word problems

This can take two forms. Some students will scan a problem to see if they can quickly spot the operation that is being asked for. Often students have been taught to approach problems in this way by a previous teacher, who has encouraged students to scan and highlight key words and then use this information to quickly figure out what operation they need to perform. These students might put their hands up and ask a question such as, "Is this question asking us to divide?" Other students may carry out the first operation within a problem and then stop. These students have become used to problems having a single answer each.

Tony was working with a group of 10- and 11-year-old students on solving word problems. They were given a table of information which told them a serving of cereal contained 350 calories per 100g. They were asked to calculate how many 30g bowls of cereal they would need to eat to meet the daily requirement of 2,000 calories. Many of the students realised that the operation they needed to carry out was to divide but did not take care when they were deciding what to divide. Some also were looking for an answer that was exactly 2,000, not realising that it would be OK to eat more than 2,000 calories. So, for example they answered:

5.714285714 (the answer on a calculator if you divide 2,000 by 350).

It would be difficult to eat such a precise amount of cereal, and five bowls of cereal would not really satisfy a growing child over a whole day.

67 (the answer if you divide 2,000 by 30 and then round up to a "whole" number of bowls).

This is quite a lot of cereal!

There are three ways to support students in not just jumping to the first operation they see. First, encourage students to always check if their answer is reasonable by linking it ack to the context of the original problem. This does mean that you need to make sure the contexts are mathematically accurate. So, for example, if you are using prices, try to be accurate. Students will have an understanding of the real costs of certain items and will be able to spot when you are just making things simple.

A useful strategy is to give students the answer to a multi-step problem and ask them to explain why this is correct. So, in this case, give the students the same information but tell them that:

19 bowls of cereal provide you with 1995 calories, which is very close to the daily requirement

and ask them to explain why this is correct. Another useful method is to present students with mathematical data and ask them to set each other multi-step problems, together with a solution. Getting students to pose their own questions is a very good way of teaching them to solve problems. This helps them to get inside the process of problem posing, and once they begin to understand how you set multi-step problems they become more adept at solving them.

Teaching point 2: not using a systematic approach to solve a problem

This area of difficulty is directly linked to the first one often arises from students thinking that the result of a problem-solving activity is a simple answer that the teacher knows and they have to find out as quickly as possible. For example, a group of 9- and 10-year-old students were working on the following problem:

How many different ways can you arrange four different coloured cubes?

Say we are using red (R), blue (B), green (G) and yellow (Y) cubes. One way would be:

Another would be:

The group started to guess. "Twelve," one group said, "because it's 4 × 3." Another group suggested 24, although they had no reason. They were just taking a non-committal response to mean that 12 was incorrect. So, the groups were given boxes of different coloured cubes and asked to find all the different arrangements. A group of pre-service teachers were asked to explore how many ways they could arrange four different coloured cubes in a line so that each arrangement is different. They used sketches to make sure they could convince each other that they had found them all. One group came up with this solution:

Their description was as follows:

Here I have made sure that I have found all the arrangements that start with a red cube. I also began by realising that if I place red, then yellow, there are two ways to arrange the blue and green cubes, and that this pattern repeats.

This is a good example of articulating mathematical reasoning, and it can be seen as a proof because it is convincing.

An effective way to encourage students to work systematically is to make sure the problem is complex enough that a system will be needed to keep track of all the possibilities. The problem we have been working on here demands a system as it is then easy to see the patterns. The most important question you can ask students to encourage them to work systematically is, "Are you sure you have found all the solutions? How do you know that you have found them all?"

Teaching point 3: difficulty explaining the thinking process ("I just did it")

One of the big ideas introduced at the beginning of the chapter was the ability to communicate mathematical reasoning. By the time students are leaving primary school we would hope they will be using sophisticated mathematical language to describe mathematical reasoning as mentioned in the progression section. However, this is not something that comes naturally. You will probably have asked a student how they have worked something only for them to tell you, "I just did it."

This is not through an inability to describe their thought process. Until we start to try to remember how we are thinking mathematically it really does feel as though we "just did it." One way to get students to describe their thinking is to present a mental calculation. For example, working out how many students are in school today.

If the number of students in a school are:

Year 1	28	Year 4	22
Year 2	31	Year 5	25
Year 3	25	Year 6	30

Can I ask you to calculate the total number in school? Try to do this mentally. Now write down how you carried out the calculation. Two 11-year-old children who worked on this said:

> I added all the tens, so 20 + 30 + 20 + 20 + 20 + 30 and that gave me 140, then I added on 21 because that is what the units came to.
>
> I noticed that 28 and 22 made 50, and that 25 and 25 made 50 so that was 100, then I added on 31 and 30.

The first student was prompted to think more carefully about how they had worked it out when they were asked to explain how they added the "tens." When they thought about this, they realised they had added all the "20s" together first, which gave them 80. They then doubled 30 to get 60, which gave them 40. The best way to help students explain their thinking is through your questioning. Look at this example.

DOMINO TOTALS

Play a game of dominoes. Can you make a total of 6 each time? Can you use all the dominoes? Play again. Try to make totals of 5. Can you use all the dominoes? Try for other totals.

What do you notice?

You will see that the text offers open questions, such as, "Can you use all the dominoes?" and asks the students to write about what they notice. Another question invites them to try other totals. You might want to support students in getting started on this by thinking about which totals are possible and which are not possible. Try to respond to an answer from one of your learners with a question. For example, if they say 13 is not possible, ask why, even though it seems obvious. You might want to encourage different pairs to try different totals too. Some will be easier than others and the pairs will then have an outcome to talk about.

Teaching point 4: unwillingness to try to prove an assertion ("It just works")

A group of Year 2 students are making patterns with hexagons and equilateral triangles.

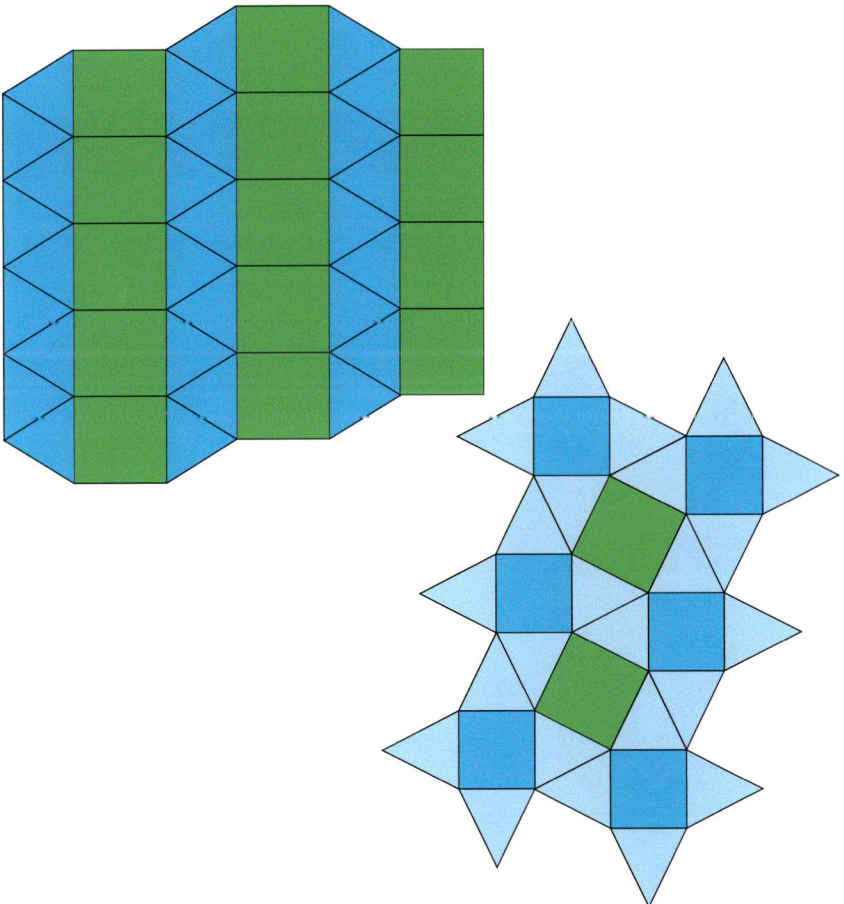

Their teacher is trying to work with them to get them to "prove" some of their findings:

Teacher: What if you made two rows of squares around your design, how many triangles will you need at the corner?

Child: Two, I think. Oh no, look, it takes three.

Teacher: And what if you made three rows of squares?

Child: I don't want another row of squares.

Many students do not immediately develop an inclination to solve problem. We can only work to develop the willingness of our students to prove their assertions by modelling this expectation as the norm in our classrooms. This comes down to questioning. Several students in classes we have taught have ended up frustrated with our usual response to an assertion they offer (e.g. "Why?" or, "Are you sure?") However, they quickly realise that being sure of an answer is a measure of success and they stop simply saying the answer and begin to explain how they know it is the right answer.

Developing ideas of proof can be structured carefully. Look at this example in the following box:

PENTOMINOES

- How many different ways can you arrange five squares so that two sides touch exactly?

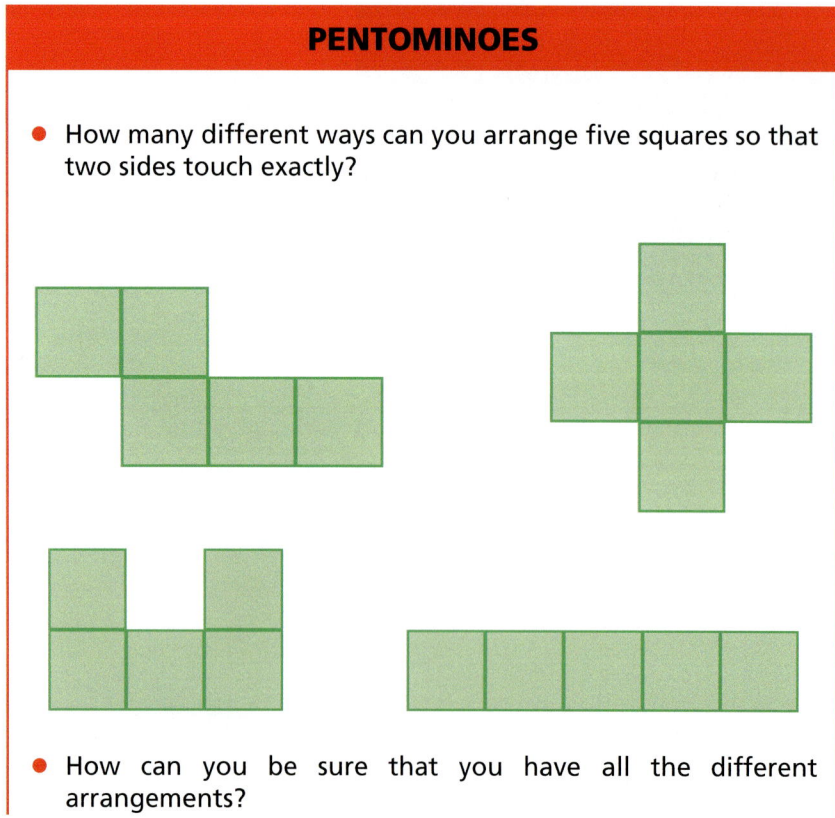

- How can you be sure that you have all the different arrangements?

- Sketch all your arrangements on squared paper. How many of these can you fold up to make an open cube?

- Is there a rule you can write down that describes all the shapes that are nets of an open cube?

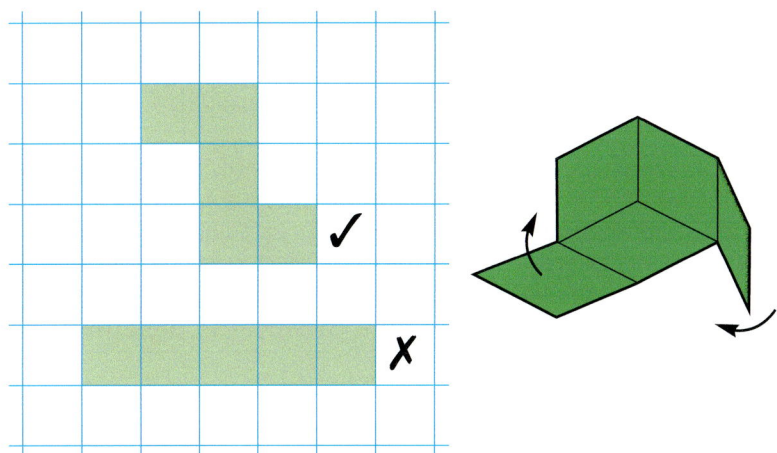

Try the activity for yourself. The activity is open at first, but by asking you to check that you have all possibilities we move you towards proof. You have to introduce some system to your recording so that you can check that you have all possibilities. Giving a convincing reason why something is the case is the beginning of proof. We then ask you to think about the shapes you have drawn in a different context. This is one way of asking you to make connections between different areas of mathematics. This time you can check your conjectures by cutting out the shapes and seeing if they will make an open cube. Then we move to trying to come up with a rule, which allows you to generalise from these special cases.

Teaching point 5: not applying logical thinking

You probably find your students will look back over correct answers they have written down and calling you over to ask you if they are correct. Try asking them what they think. Maybe they will say something like, "Well, I think it's right but it seemed too obvious." It is almost as if they expect mathematics to trick them. If something seems obvious or logical, they must have got it wrong. Mathematics cannot be that easy. This may be because they have just been taught a series of tricks and routines rather than having been shown that mathematics is inherently logical and follows very predictable patterns. If I can add

7 to 15, I can follow exactly the same process to add 7 to anything. And any time I add 7 to a number with 5 ones my answer will have a 2 in the ones column.

One of our favourite problems to work on with students to help them see the importance of logical thinking and careful recording is the chessboard problem. We would ask you to work on this too. The question is very simple:

How many squares are there on a chess board?

At first you will think the answer is straightforward. There are 64 squares on a chessboard. But if you look carefully there are many 2 × 2 squares, many 3 × 3 squares and so on. Right the way up to an 8 × 8 square which forms the whole board. The next task is to find a way of counting these different squares. We will leave you to work on this.

How many squares are there on a chessboard? The answer is not 64! The answer is not 65!

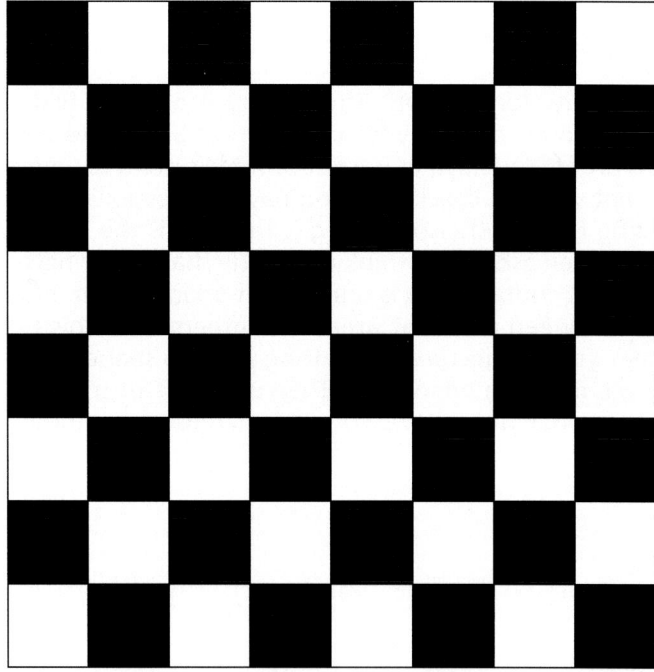

When you have an answer can you use this answer to work out how many squares there would be on a 10 × 10 board?

For those of you that are interested in teaching logical thinking the Association of Teachers of Mathematics (ATM) in the UK have published a series of books designed to support learners develop logical thinking and problem solving skills. Each book contains a series of activities presented a set of cards. Each set of cards poses a problem and contains all the information needed to solve it. The set also contain information that is correct but of no relevance or use in solving the problem.

Teachers who took part in trialling of the activities reported that children began to apply some of the skills and behaviours they exhibited when working with the cards to other work. In particular, they reported a more logical and strategic approach to organising information and a greater inclination to check answers. This is more likely to happen when there is an explicit analysis of the approach that was followed to reach a solution.

The books draw on a range of different areas of mathematics including number, shape, time, data handling and money as well as focussing on more general thinking skills such as logical deduction and the organisation of information.

A number of benefits were noted by teachers in using these activities, including the following:

- they are motivational
- groups with a wide range of attainment can be active participants
- they put problem solving into meaningful or interesting contexts
- they teach children not to expect immediate answers to problems
- they make multi-step problems more manageable and in so doing encourage a strategic approach
- they encourage children to organise information and identify redundant information
- they develop skills of collaboration and cooperation
- they encourage children to check their answers.

The following example can be used for a collaborative problem-solving activity exploring the arrangement of shapes:

Discover the colour of each shape and how they have been arranged.	All the shapes with straight sides are regular shapes.	There are two rows of four shapes.
Yellow and blue shapes are next to each other in the top row.	The shapes in the top row are the same as the shapes in the bottom row but in the reverse order.	The blue shape has no straight lines.
There are four different shapes in the arrangements.	The shape with the largest number of sides is green.	The hexagon on the bottom row is below a triangle.

The first shape in the top row is half the area of the second shape.	Each different shape has its own colour.	The number pf letters in the colour of the shape with the smallest number of sides is the same as the number of sides.
One of ten shapes has four sides that are equal.	What are the names of the shapes in the arrangement?	The blue shape has a diameter that is equal to the length of one side of the yellow shape.

Paul Swan and David Dunstan have developed a similar set of cooperative problem-solving activities in their book, *Check the Clues*. These are available from www.drpaulswan.com.au. Each book in the series is pitched at a specific year and the expectation is that they will be used with all students to develop their problem-solving, logical-thinking and collaboration skills.

In practice

This lesson plan was used with a group of Year 3 and 4 students. The problem is built around a multiplicative situation that many students do not immediately recognise as one they can solve with multiplication. This is because it explores how many combinations are possible, rather than presenting a "groups of" situation that most students can easily connect with multiplication. Following the plan is an evaluation of the lesson that explores how successful the plan was in supporting the children to develop their knowledge and understanding.

Learning intentions

To explore a real-world multiplicative situation involving combinations.
To connect concrete and pictorial representations of the problem with matching repeated addition and multiplication equations.

Key vocabulary

array, Cartesian product, for each, times, multiply.

Context for lesson

This lesson was taken mid-way through a unit of work focussing on multiplicative thinking. Previous problem-solving lessons indicate that students are confident solving "groups of" problems and representing their thinking in several ways, including diagrams, models with concrete materials, tables and equations. This was the class's first exposure to a combinations problem.

Warm-up activity: date maths

Ask the students to work in groups to try and make an equation for each of the numbers from 1 to 10 using only the digits in that day's date. For example, the 13th of June 2021 would be recorded as 13/06/21, giving the students access to the digits 1, 3, 0, 6, 2, and a second 1. Students can combine the digits in any order to make larger numbers, but you can only use each digit as many times as it appears in the date. So, in this instance, 31–21 would be a perfectly fine equation for 10, but 22 ÷ 11 would not be accepted for 2, as the student has used the digit 2 twice, despite their only being one 2 in the date. To encourage participation from a broad range of students, each person can only provide one equation.

Launch

I told class that I visited the local cinema to see the latest instalment of the Star Wars movies. Then I read the text version of the problem to the class.

At the Sun Theatre, they sell choc top ice-creams where you can choose among having a white, milk or dark chocolate topping. They also offer a choice of seven different flavours: strawberry, chocolate, vanilla, banana, mint, rainbow and berry swirl.

If each ice-cream combines one flavour with one type of chocolate topping, how many different types of ice-creams can you choose among?

This was followed by a brief discussion where any unfamiliar terms in the problem were clarified for the class. The expectations for the first five minutes of the explore time were also clearly outlined. After this, the students were sent off to begin tackling the task. Notably, there was no modelling of solutions during the launch of the problem.

Explore

I asked students to work independently on the task in question in complete silence for the first five minutes. After this initial silent period ended, students were actively encouraged to collaborate, allowing them to share their ideas and compare different methods they are using. During the explore time, the following enabling prompt was made available for students who had trouble getting started.

Enabling Prompt

If Michael selected white chocolate, how many different ice creams could he make? Use a diagram to record your answer.

There was also an extending prompt given to students who had finished the existing task.

Extending Prompt

Another cinema sells a different range of ice-creams and they have 48 different combinations in total. How many flavour and topping combinations might they have for sale to arrive at a total of 48 choices? Find as many solutions as you can.

Summary

At the end of the lesson, students returned to the floor to reflect on the lesson. In this instance, the teacher invited two students to the whiteboard to share their approach to the task. Molly solved the problem by systematically drawing diagrams. She started by focusing on one flavour (strawberry) and then drawing a simple diagram for each of the three possibilities (white, milk and dark). She repeated this process for all 7 flavours and then counted her diagrams to discover 21 possible choices.

Next, Cale was asked to share his thinking. Cale found the same solution but used a different approach. He showed the class how he drew up a table with three columns, one for each type of chocolate topping. Cale then completed the first column by filling in the seven flavours. He did the same for the second column before realising that he could work the answer out by using a multiplication equation of 3×7. Once Cale finished his explanation, the teacher then drew the class's attention to the similarities between Molly and Cale's work.

Rationale and evaluation

The lesson was successful, as all students were engaged with the task. During the warm-up, the teacher made a conscious effort to promote creativity by posing questions like, "I wonder if anyone can make an equation for one of the numbers using all six digits?" or "Can anyone make an equation using both division and addition?" He also praised students who pushed themselves to make more challenging equations.

Around three-quarters of the class arrived at the solution of 21 for the main problem that was given to the class, but how they got there significantly varied. Some students used the enabling prompt to get them started; others worked it using models, diagrams, tables and equations. All the students who did not arrive at the solution were well on track to figuring it out, with quite a few finding many of the possible solutions but not working systematically. The summary at the end of the lesson was beneficial for these students, as Molly's systematic approach to drawing her diagrams clearly illustrated an excellent way to tackle this type of question.

Six or seven students made significant progress with the extending prompt, finding many of the available answers. There was a group of students who were working together, and they were on track to finding all of the possible solutions, given that they were working through all of the possible combinations in order. Ellie said, "We've got 1 x 48; 2 x 24; 3 x 16; 4 x 12 then 5 x er, nothing as this won't work."

Assessing problem solving using mathematics

When you wish to assess your learners to ascertain their skills in problem solving using mathematics you need to make sure you give your learners the opportunity to evidence: how they are approaching the problem that you have set them; the extent to which they are being systematic; their abilities in predicting possible outcomes and testing these; the effectiveness of the recording systems that they adopt and finally the effectiveness of their communication of their reasoning process to others. As an example, reflect back on portfolio task 3.1 and answer the following questions as honestly as you can:

1   How persistent were you at looking for a range of patterns? Did you explore a range of patterns or stop after you had found two or three?

2   Were you able to explain why the pattern appeared, or did you just describe the pattern?

3   Did you use "increasingly complex" mathematical vocabulary to describe these patterns?

4   How systematic were you in exploring whether every number appeared in your search for 1,000?

5   Did you have a convincing explanation for your answer to whether or where 1,000 appeared?

Try this activity with the group you are currently teaching, or try any of the other activities in this chapter, and use the previous questions to assess your pupils. Be open with them about the criteria you are using to assess them so that they are clear about your expectations. It is important that your pupils become aware of what counts as "success" in problem solving as this may be slightly different to previous success criteria.

Cross-curricular teaching of problem solving

Students respond very well to solving problems in their real world. This can also help them see the point of learning new skills and motivate them to come to new understandings. A cross-curricular project offers you the possibility of working with mathematics over extended periods of time rather than in small chunks, often only an hour. Of course, when we problem solve outside school, we don't limit ourselves to a single hour a day in order to solve our problem.

A project focusing on developing a more sustainable school will both engage your learners and allow you to engage in a wide range of problem-solving activities. The most effective way to begin such a project is to ask your learners how they could explore this issue. Do some research for yourself first. How much wastepaper is thrown away each day? How much does the school spend on paper and photocopying? What are the annual fuel bills? What are the implications for pollution for the ways that pupils travel to school? Some of the areas that students have explored on this extended project over the years have included the following:

- **Car parking:** what is the most effective way of redesigning the car park so that it uses a minimum of space and so that it can incorporate a bike shed to encourage cycling to school. This involves measurements of cars, the construction of scale models and trying out different arrangements for parking spaces. The issue of ensuring there are sufficient parking spaces for those needing accessible parking is also important

- **Travel to school:** Students always enjoy surveying the whole school to find out how teachers and students get to school. They can then calculate the current carbon footprint. This has led to students exploring ways of lift sharing, of walking buses, and of cycling schemes where students meet up and cycle to school together. The mathematics involves exploring shortest routes to school and possible "pick-up" points.

- **Recycling paper:** Again, a brief survey can ascertain which classes use most paper; which classes are best at recycling paper; and the money that can be saved through recycling paper internally as scrap paper or in home-made exercise books. A trip to the recycling plant is always popular.

- **Energy bills:** Sharing the energy costs within a school has always led to students becoming incredibly vigilant about turning off lights and thinking carefully about energy usage.

Summary

The focus of this chapter has been problem solving using mathematics. This chapter is deliberately placed early in the book as the whole approach of the book in developing your subject knowledge is through a problem-solving or investigative approach. We hope that the chapter has given you insight into how you can learn mathematics through problem solving and that this in turn will enable you to teach children to problem solve and investigate effectively. The key ideas of **communication**, **reasoning** and **enquiry and problem solving** were outlined early in the chapter as a set of key skills in this area. The big ideas section also introduced you to ways in which **specialising** and **generalising** underpin any mathematical problem solving. The portfolio tasks throughout the chapter and the exemplar activities within the teaching points will also give you a wide range of starting points with which you can explore mathematical problem solving with the children you teach.

Reflections on this chapter

There are several key points to take from this chapter. First, you should now feel able to make a start on an investigation that is open and you should be able to support your learners in attacking an investigation through questioning rather than through directing them too closely. Perhaps you will be able to work with the children you teach to explore mathematical problems with them so that you model being a mathematician. You should also have an understanding of how important it is to work with very young children on describing their thinking processes and realise that you can teach children to articulate their thoughts. You will have become more aware of the way you think mathematically when working at problem-solving activities. You can then share your thinking processes with the students you are teaching as a way of helping them understand their own thinking processes.

You have been introduced to key concepts such as specialising and generalising. Just a reminder: specialising means to look at specific examples to look for patterns; generalising means being able to make a statement about all cases. The most important thing that you should take away from this chapter is that a problem-solving approach means teaching your students to act like mathematicians rather than to simply follow rules that you have given them. If the students you are teaching are reflecting on their thinking, asking interesting questions about the problems they are working on and most importantly "having an inclination" to solve problems, you will be succeeding.

Going further

The following books will help you explore the area of problem solving using mathematics in much more detail. The books also offer a wide range of activities both for you and your students.

Supporting teachers in structuring mathematics lessons involving challenging tasks by Peter Sullivan and colleagues

This report summarises an investigation that explored ways of supporting teachers to convert challenging tasks into classroom lessons, while also supporting students to engage with these tasks. It covers the important role that lesson structure plays in reducing anticipated negative responses from the students, as well as examining the role of enabling and extending prompts to cater to differences in student performance.

Sullivan, P., Askew, M., Cheeseman, J., Clarke, D., Mornane, A., Roche, A. and Walker, N. (2015) 'Supporting teachers in structuring mathematics lessons involving challenging tasks', *Journal of Mathematics Teacher Education*, 18(2), 123–140.

Using prompts to empower learners: exploring primary students' attitudes towards enabling prompts when learning mathematics through problem solving by James Russo and colleagues

This paper examines how enabling prompts can be used to assist teachers with differentiation when designing problem solving tasks, exploring their use from the students' perspective. The results of this study found that having access to enabling prompts led to the students feeling empowered, due in part to increased levels of student autonomy, thus allowing them to take more responsibility for their learning. Students also felt the prompts increased their levels of understanding, allowing them to approach the task in question with more confidence.

Russo, J., Minas, M., Hewish, T. and McCosh, J. (2020) 'Using prompts to empower learners: Exploring primary students' attitudes towards enabling prompts when learning mathematics through problem solving', *Mathematics Teacher Education & Development*, 22(1), 48–67.

The elephant in the classroom: helping children learn and love maths

JO BOALER

This book draws on Jo's research in both the UK and the USA exploring how children best learn maths. She has followed thousands of learners over several years and argues convincingly that problem solving through mathematics should be at the centre of the curriculum. The book includes examples of problem-solving activities that she has used or observed. You should also visit her "youcubed" website www.youcubed.org, which offers lots of activities you can use in your classroom. There are also several papers describing Jo's research in this area. The tagline is "depth not speed."

Boaler, J. (2015) *The Elephant in the Classroom: Helping Children Learn and Love*. London: Souvenir Press.

Thinking mathematically

JOHN MASON, LEONE BURTON AND KAYE STACEY

In *Thinking Mathematically* the authors carefully describe the "art" of problem solving that lies at the heart of mathematics. They demonstrate how you can encourage, develop and foster this process both in your own thinking and in the students that you teach. They do this by carefully taking you through a series of problem solving activities.

Mason, J., Burton, L. and Stacey, K. (1985) *Thinking Mathematically*. London: Pearson Education.

KAYE STACEY

In *What is mathematical thinking and why is it important*, Kaye Stacey explains why she sees problem-solving as at the heart of teaching and learning mathematics. She argues that problem solving involves bringing together content knowledge and problem solving skills. She also argues that teachers should model the problem solving process and engage in mathematical thinking themselves throughout any lesson.

Stacey, K. (2006) *What Is Mathematical Thinking and Why Is It Important*. Available at file:///Users/home2/Downloads/WHAT_IS_MATHEMATICAL_THINKING_AND_WHY_IS_IT_IMPORT.pdf

Audit task

For every chapter you will be asked to write your own lesson plan exploring an area covered in the chapter. You should focus on learning which is appropriate for a group of students that you are working with and select an area of mathematics that has been discussed in this chapter. Construct a lesson plan using the proforma on the companion website or any planning proforma that is used by your course or by the school in which you work. Teach the lesson and then evaluate it carefully with a focus on student's learning and misconceptions.

Add this lesson plan and evaluation to your subject knowledge portfolio. This is a very important piece of evidence to show your developing subject knowledge.

Self-audit

You should carry out this self-audit and add it to your subject knowledge portfolio. This offers evidence of your own learning and development in the area of "problem solving using mathematics." Read through the problem several times before you try to make a start. Also reread the description of rubric writing earlier in the chapter so that you can use this process to describe your thinking.

This task explores number patterns within a 100 square:

1	2	3	4	5	6	7	8	9	10
11	12	13	14	15	16	17	18	19	20
21	22	23	24	25	26	27	28	29	30
31	32	33	34	35	36	37	38	39	40
41	42	43	44	45	46	47	48	49	50
51	52	53	54	55	56	57	58	59	60
61	62	63	64	65	66	67	68	69	70
71	72	73	74	75	76	77	78	79	80
81	82	83	84	85	86	87	88	89	90
91	92	93	94	95	96	97	98	99	100

Look at a block of four numbers, for example:

15	16
25	26

If I add these numbers together I get a total of 82.

This is (4 × 15) + 22, as 4 × 15 = 60 and 60 + 22 = 82.

Picking another block and looking at the number in the top left box

18	19
28	29

I notice that (4 × 18) + 22 = 72 + 22 = 94 and 18 + 19 + 28 + 29 = 94 too.

This is an example of specialising – I have looked at two different blocks of four squares and explored the patterns in them. In order to generalise I can write down

n	$n + 1$
$n + 10$	$n + 11$

or

n	One more than the number to its left
10 more than the number above	11 more than the number in the top left square

So, adding these "numbers" together:

$$n + (n + 1) + (n + 10) \ 1 \ (n + 11) = 4n + 22$$

This shows that if I take any section of four numbers from the grid and add them together I will always get a total that is four times the number in the top left corner + 22.

Now look at other blocks of numbers in the 100 square on page 51, for example:

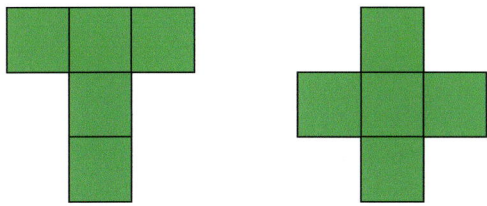

As you work at these problems jot down your thinking process at the side of the page. Use the rubric writing mentioned in the chapter. How do you get started on the problem? When are you stuck? What questions do you ask yourself to move forward? When are you specialising and when are you generalising? What "aha" moments do you have? And finally, what have you learnt about this problem and yourself as a mathematical thinker?

Answer to logic problem

Red triangle	Yellow square	Blue circle	Green Hexagon
Green Hexagon	Blue circle	Yellow square	Red triangle

CHAPTER 4

NUMBER
Counting; place value; fractions, decimals and percentages

Introduction

It is very difficult trying to remember how we learnt to count. It is one of those things many of us learnt before we had clear memories. However, you have probably either worked with or been with young children beginning to make sense of counting. Tony blogs regularly about observing his grandsons, currently aged eight and four, making their first steps in learning about numbers and the numbers system. He remembers sitting with a three-year-old child of a good friend who had learnt to count to 20. He did not have a sense of what "counting to ten" meant but he had heard the pattern repeated many times. He counted carefully, saying each word accurately, "one, two, three. . . ." and so on. I wondered what would happen if I deliberately miscounted and so I jumped from six to eight missing out seven. For some reason he thought this was hilarious. "No," he shouted at me, so of course I repeated the joke. He did not yet know that eight was two more than six, but he did know that you should always say seven immediately after six.

This chapter will unpack the big ideas in learning to count and in understanding number. It will also engage you in activities that will allow you to reflect on how you learn counting and understanding number and will draw on research to illustrate how best to support children in their learning. This opening activity helps you reflect on the skills you bring to counting without really thinking about it.

Starting point

Portfolio task 4.1

You will need to work with a small group of friends or colleagues.

Take it in turns to pick up a number of small cubes, between 5 and 12. Drop them onto a flat surface and cover them with a piece of cloth. Remove the cloth so that your friends can see the cubes but replace it before they have chance to count the cubes.

Ask them how many cubes they saw.

Then ask how they calculated. It is likely that the group will have counted in twos or threes or even larger groups depending on the arrangement of the cubes.

Try this several times with increasing numbers of cubes to see how many cubes they can "count" in this way.

This activity illustrates the basic principles of counting, which will be explored in more detail later in the chapter. The skill of being able to see how many objects there are, without physically counting each one, is called subitising. When children subitise, they are able to see objects in a range of ways to help them count quickly and understand that the total of the count is the same as the total of the groups they are counting. They probably have a range of mental images attached to particular numbers. For example, they might see five as an arrangement of dots on a die, as two dots next to three dots, or as a line of five dots. This means that when they see the number "5" they also see:

5 = 2 + 1 + 2   5 = 3 + 2

Some children with additional learning needs, for example children who have dyscalculia, may have difficulties subitising. It is a helpful skill to teach as it allows us to see and partition numbers in different ways, as in the previous example.

It can be argued that counting and understanding place value is at the heart of mathematics: the essence of numeracy. There are many counting games and traditional rhymes which show that counting is a skill that is often developed outside the traditional classroom. There are also different traditions for finger counting; some cultures count each finger separately, some use knuckles so that each finger can represent a count of three, others use the thumb as a symbol for five or ten. There is a myth that Aboriginal and Torres Strait Islander peoples do not count beyond five. That the count is basically one, two, three, four, five, many. An article by Shannon Foster for the University of Sydney explains that most of this theory arose from linguists who did not understand the cultural underpinnings of Aboriginal peoples and Torres Strait Islanders' expression of numbers. She suggests that we will still see Aboriginal and Torres Strait Islander peoples using body tallying to count. The *Wurundjeri* counting system related directly to body parts.

Aboriginal name	Literal translation	Translation	Number
Giti mŭnya	little hand	little finger	1
Gaiŭp mŭnya	from *gaiŭp* = one, *mŭnya* = hand	the ring finger	2
Marŭng mŭnya	from *marung* = the desert pine *(Callitris verrucosa)*. (i.e., the middle finger being longer than the others, as the desert pine is taller than other trees in Wotjo country)	the middle finger	3
Yolop-yolop mŭnya	from *yolop* = to point or aim	index finger	4
Bap mŭnya	from *Bap* = mother	the thumb	5
Dart gŭr	from *dart* = a hollow, and *gur* = the forearm	the inside of the elbow joint	6
Boibŭn	a small swelling (i.e., the swelling of the flexor muscles of the forearm)	the forearm	7
Bun-darti	a hollow, referring to the hollow of the inside of the elbow joint	inside of elbow	8
Gengen dartchŭk	from *gengen* = to tie, and *dartchuk* = the upper arm. This name is given also to the armlet of possum pelt, which is worn around the upper arm.	the biceps	9
Borporŭng		the point of the shoulder	10

Aboriginal name	Literal translation	Translation	Number
Jarak-gourn	from *jarak* = reed, and *gourn* = neck, (i.e. is, the place where the reed necklace is worn)	throat	11
Nerŭp wrembŭl	from *nerŭp* = the butt or base of anything, and *wrembŭl*= ear	earlobe	12
Wŭrt wrembŭl"	from *wŭrt* = above and also behind, and *wrembŭl* = ear	that part just above and behind the ear	13
Doke doke	from *doka* = to move		14
Det det	hard	crown of the head	15

The big ideas that are explored in this chapter are counting; place value; fractions, decimals and percentages and proportionality. The next section shows the progression in counting and understanding number in terms of the key objectives within the framework.

Progression in counting; place value; fractions, decimals, percentages and proportionality

Don't worry if you don't understand all of the ideas in this section at the moment. You will have been introduced to the key vocabulary by the time you have finished working through this chapter. All of the ideas are discussed in detail as you work through the chapter.

Foundations for counting and understanding place value:

Initially, young learners will learn to say and use number names in order. They will start to understand that numbers identify how many objects are in a set and use this understanding to count reliably up to ten everyday objects. They will be able to subitise (see later in the chapter) small numbers of objects (the activity that opened this chapter would be a useful assessment activity to see how well this skill has been developed). They will be able to compare collections of up to 20 objects and explain their thinking.

Beginning to count and understand place value:

The next stage is for students to read and write numerals up to initially to 100 and later to 1,000. They will skip count in twos, fives and tens from zero. They will use this skill to explore number sequences.

They will also learn to explain what each digit in a two-digit number represents, including numbers where zero is a place holder. ("0" is often described as a "place holder" as it has no value. So, for example, in the number 4,026 there are no "hundreds," but we need to use the "0" as otherwise we would confuse 4,026 with 426.) When they understand the meaning of the digits in two-digit numbers they can be taught about **partitioning** two-digit numbers in different ways, including into multiples of 10 and 1. Partitioning is a very useful skill to develop in order to support the development of mental methods. For example, we can partition 47 into 40 + 7. So, if we are to try to add 47 + 15 we can see it as 40 + 7 + 10 + 5 or 40 + 10 + 7 + 5 = 62. Students will be beginning to understand the idea of halves, quarters and eighths of shapes and quantities, knowing that there are two halves in a whole and four quarters in a whole.

Becoming confident in counting and place value and developing an understanding of fractions, decimals and percentages

Students can now develop their understanding of partitioning so that they can partition four-digit numbers into multiples of 10,000, 1,000, 100, 10 and 1 in different ways. They can investigate number sequences involving multiples of 3, 4, 6, 7, 8 and 9. They will also investigate the properties of odd and even numbers.

They are also ready to model and represent **unit fractions** including $\frac{1}{2}, \frac{1}{4}, \frac{1}{3}, \frac{1}{5}$ and their multiples to a complete whole. They will also investigate equivalent fractions in familiar contexts and count by halves, thirds and quarters locating these fractions on a number line.

They can also explain what each digit represents in whole numbers and decimals with up to two places, and partition and order these numbers.

Extending learning in counting and place value and developing an understanding of fractions, decimals and percentages:

Towards the end of their time in primary school students will identify and describe factors and multiples of whole numbers and use this knowledge to solve problems. They will be estimating and rounding to check the reasonableness of an answer.

Their knowledge of fractions and decimals will allow them to compare and order **unit fractions** and fractions with related denominators. They will also know that place value can be extended beyond hundredths, and they will use this knowledge to compare, order and represent decimals. They will be making connections among equivalent fractions, decimals and percentages.

The next section of the chapter focuses on the big ideas of counting; place value; fractions, decimals and percentages and proportionality, which are at the heart of counting and understanding number. If you are confident with these ideas, you will be able to teach children successfully across the primary age range and will have a clear understanding of the development of these ideas across the 4–12 age phase.

Big ideas

Counting

Many children will start school able to count, but it is important to be aware of the principles of counting, both to support the children who cannot yet count and to recognise the processes that young children who have already learnt to count have mastered. The key research that supports us in understanding the process of learning to count was undertaken by Rochel Gelman and Randy Gallistel in *The Child's Understanding of Number* published in 1986. Gelman and Gallistel are both psychologists, and their book was seen as marking a huge development in our understanding of how children learn to count. Through careful observation of young children undertaking activities that they had planned, they described five principles that underpin counting:

1 **the one-to-one principle:** a child who understands the one-to-one principle knows that we say one number for each object and we only count each item once

2 **the stable order principle:** a child who understands the stable order principle knows that the order of number names always stays the same. If we are counting by ones, we always say "one, two, three, four, five . . ." in that order

3 **the cardinal principle:** a child who understands the cardinal principle knows that the number they attach to the last object they count gives the answer to the question "How many . . . ?"

4 **the abstraction principle:** a child who understands the abstraction principle knows that we can count anything – they do not all need to be the same type of object. So we can count apples, we can count oranges or we could count them all together and count fruit

5 **the order irrelevance principle:** a child who understands the order irrelevance principle knows that we can count a group of objects in any order and in any arrangement and we will still get the same number.

Portfolio task 4.2

Can you think of examples of children who do not yet understand these principles?

For example, a child who is asked to count seven multilink cubes laid out in a row and then count them again when you put them into a different arrangement to check there are still seven would not yet understand the order irrelevance principle.

Write your examples and enter this reflection in your personal portfolio.

The mathematics curriculum in New Zealand has recently been developed and they have published a series of progression guides that support teachers in planning for progression at the early stages of children's learning. They break down teaching to count into the following steps:

1 **pre-counting:** the focus here is on the ideas of more than, less than and the same as. Children will often make these comparisons by lining up objects rather than counting

2 **one-to-one counting:** this is the ability to say the number words in order and to match each number to one object as described earlier

3 **counting sets:** this links to the cardinal principle and the stable order principle

4 **counting from 1 to solve number problems:** these may be addition problems involving combining sets and counting all the objects in both sets, or they may just be questions like "how many . . . ?" Children can also count from one to solve subtraction problems. They count from one to see how many objects to remove from a set and then count how many are left.

5 **counting on to solve number problems:** by this stage children understand the cardinality principle and can count forwards and backwards and so can add seven and two by starting counting at seven.

Concrete, pictorial, abstract: the CPA approach

The ideas of Martin Hughes were developed from Jerome Bruner's three modes of representation: enactive; iconic and symbolic. Bruner suggested that in the enactive phase, children will draw the objects they are using to count. So, if they are counting cubes, they will draw cubes. If they are counting cars, they will draw cars. The system of representation develops to become iconic. Here children might use tally marks or simple shapes to represent the objects they are counting. Finally, children will start to use symbols that are recognisable as numbers to represent the count.

Recently there has been an increased emphasis on the use of manipulatives to support children's understanding of place value. This uses the learning theories of Bruner and Hughes to develop a teaching approach. Here, manipulatives such as icy pole sticks and elastic bands or counters and tens frames are used to model the place value system before eventually moving to base ten blocks. Base ten blocks are not used initially, as many students find it difficult to see the key feature of place value, that the bundling of ten ones can represent one ten. Thus, extensive experience with physically bundling concrete materials to represent numbers is an important step prior to introducing base ten blocks.

Thousands	Hundreds	Tens	Units/Ones
1000	100	10	1

This can be seen as an enactive representation if a number. The next step is to move to using place value counters. With place value counters a single counter represents a one; a counter of a different colour represents a ten and, similarly, another counter represents a hundred. So, for example, 332 can be represented as
It is then a short step to the symbolic representation using a place value grid.

Hundreds	Tens	Ones
3	3	2

TAKING IT FURTHER: **FROM THE RESEARCH**

Linda Marshall and Paul Swan from Edith Cowan University carried out a research project exploring the use of manipulatives in Australian classrooms. They interviewed teachers across the pre-primary to Year 10 (ages 4–16) as well as interviewing and observing teachers. They open their paper by quoting the proverb:

I hear and I forget
I see and I remember
I do and I understand,

suggesting that this has formed a philosophy for a sometimes unmediated use of manipulatives. Their research suggests that sometimes teachers take an unquestioning approach to the use of manipulatives, noticing that if teachers do not understand how manipulatives support the development of understanding of place value then their use is likely to be ineffective. They conclude that a fourth line should be added to the proverb:

I talk about it, and I connect.

To stress the importance of teachers developing their own subject knowledge around the use of manipulatives.

Place value

Place value refers to the value of a digit which is dependent on its location within a number, such as ones, tens, hundreds. So, in 352, the place value of the 5 is "tens" and the 5 is worth 50.

The system of place value consistently in much of the Western world is described as the Hindu–Arabic method. It is based on the following key principles:

There are ten digits (0, 1, 2, 3, 4, 5, 6, 7, 8, 9). The column that a digit is placed in determines its value. A digit one place to left of another digit is worth ten times its value. Zero is used as a place holder to represent an empty column.

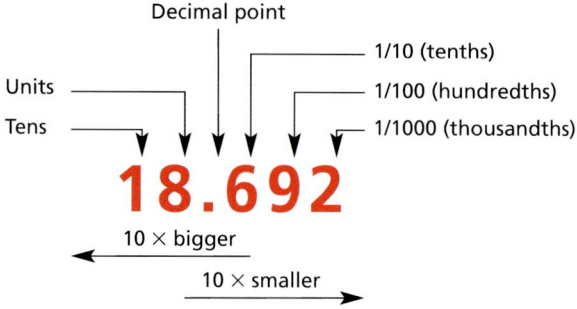

This place value system developed from the use of counting boards by traders. Beans or counters would be placed in each column on the board to represent a number. If ten beans were placed in a column they would be replaced by a single counter in the column to the left.

Teachers often ask me at what stage they should introduce "zero." The word *zero* derives from the Arabic word *sifr* meaning "empty," and that is why we need the symbol "0" or zero. We use zero to represent an empty column in place value, so the number

Hundreds	Tens	Ones	.	Tenths	Hundredths
3	0	7	.	6	

has 3 hundreds, 0 tens, 7 ones and 6 tenths. It also has 0 hundredths and even 0 millions, but we do not write the zeroes if they are at either end of the place value grid, otherwise every number would contain an infinite number of digits. So, to answer the question "When should we introduce zero to young children?" our answer would be that we use the word zero from the very beginning. Children will see zeroes around them as soon as they start noticing numbers. If we say "Can you see a zero?" in the same way that we would ask "Can you see a nine?" as we walk past number nine in the street, children will come to see that zero is not the same as the letter "O." They even look slightly different – "0" and "O." That is why there are two separate keys for the symbols on the keyboard that I am using to type this sentence.

We would not use them when counting, so start counting at 1. We do not spend our time looking for things that there aren't any of, do we? But whenever you have a number line rather than a number track, include a zero. And as soon as you begin to explore place value with your students you can introduce the idea of zero as a place holder.

As suggested in the previous section, many children begin their formal schooling able to count. However, it is not always the case that they link the counting to the symbols that represent numbers. In the book *Children and Number*, referred to earlier, Martin Hughes argued that children move through three stages when beginning to represent numbers. At first, they will represent their counting pictorially, so they will record five sweets by drawing five sweets; then they move on to an iconic form of recording so they might represent five sweets by drawing five sticks. Finally, they use the symbol "5" to represent the count.

In 1989 Herbert Ginsburg, from Columbia University In New York, developed some of Piaget's ideas to suggest that there are three facets to understanding place value:

1   A child can write numbers but does not make the link between the symbol and the number they are counting
2   A child can recognise when a number is written incorrectly
3   A child understands the value of each digit in a number.

Teaching students to understand partitioning is a useful way to develop their understanding of place value. If you look back to the progression section that opened this chapter, you can see that, at the early stages, children are encouraged to read and write numbers up to 20, which links to phases 1 and 2 in Ginsburg's schema. If you look back at the previous section discussing the use of manipulatives you can see how a variety of materials can be to represent partitioning in a concrete form.

Portfolio task 4.3

Can you think of examples of children who do not yet make the link between the numerical symbols and the results of a count?

We have heard children ask how to "spell" a certain number.

They might count 15 ducks in a pond and ask, "How do you spell 15?" Alternatively, they may recognise particular numbers such as their age, the house number they live at, or the bus they catch to school and not connect this to the numbers on a number line.

Add these ideas to your portfolio.

Fractions, decimals and percentages

Before you start this section, We would like you to try this Portfolio task.

Portfolio task 4.4

Take as large a sheet of paper as you can get hold of.

Write $\frac{1}{2}$ in the centre. Create a web diagram with as many different representations of $\frac{1}{2}$ as you can think of.

These might be diagrams with $\frac{1}{2}$ shaded in different ways, they might be equivalent fractions, or they might be percentages.

When you have exhausted this idea pick another fraction that is linked to $\frac{1}{2}$ in some way. You might choose $\frac{1}{4}$ (which is $\frac{1}{2}$ of a $\frac{1}{2}$). Create another family of representations for a quarter.

Carry on filling the paper with as many different fractions as you can think of.

This task shows how closely related fractions, decimals and percentages are. They are sometimes taught separately, which leads to students not making the link among the three ideas. In fact, they are just three different ways of representing numbers. It is very useful to move among the three representations as this allows us to see mathematics as connected rather than separate. It also allows us to draw on our understanding of one area of mathematics to solve problems in another area.

Fractions are "parts" of whole numbers. We need to be able to describe "parts" of whole numbers for two reasons. First, when measuring we can't be sure that a length or weight will also be a whole number. Second, when we "share" or divide numbers there are many occasions when the result of the division is not an integer, so we need a way of writing an answer that is not an integer. We call this a fraction.

Terezinha Nunes and Peter Bryant have explored for many years the ways in which children understand fractions. In 1996 Blackwell published their book called *Children Doing Mathematics*, and in this they suggest that understanding fractions is not simply a case of extending the knowledge we have of whole numbers. There are key differences. For example, a whole number can only be represented in one way: if we count three objects, we will write "3." However, there are classes of fractions. For example, $\frac{1}{4}$ is the same as $\frac{2}{8}, \frac{4}{16}$ or even 25% or 0.25. Fractions are also used for different purposes and appear to mean different things in different cases. So, if a fraction represents part of a whole, the denominator represents the number of parts into which the whole has been "cut" and the numerator represents the number of parts taken. In the fraction $\frac{5}{7}$, 5 is the numerator and 7 is the denominator.

Nunes and Bryant suggest that children have to come to an understanding of two key ideas:

- first, that for the same denominator, the larger the numerator, the larger the fraction. So $\frac{1}{7} < \frac{4}{7} < \frac{6}{7}$
- second, that for the same numerator, the larger the denominator, the smaller the fraction. So, $\frac{3}{5} < \frac{3}{7} < \frac{3}{10}$.

  For some, learning fractions is associated with "rules." For example:
- you must multiply the numerator and denominator by the same number when making equivalent fractions
- you must divide the numerator and denominator by the same number when making equivalent fractions
- when dividing fractions, you turn the second fraction "upside down" and multiply.

Portfolio task 4.5

This task is taken from the Nrich website. This is an open access website developed by Cambridge University and accessed by schools globally. The activities are excellent and take a problem solving approach to learning and teaching mathematics. The children (and their families) are usually quite excited to be working on activities from Cambridge University. There is also a facility to upload your classes responses to the website and compare them with ideas from other children all around the world.

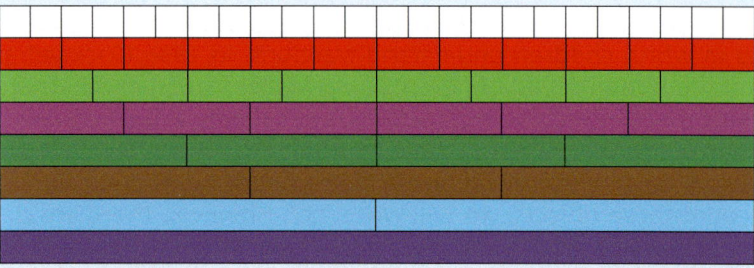

- Using the above image, how many different ways can you find of writing $\frac{1}{2}$?
- From the picture, what equivalent fractions for $\frac{1}{3}$ can you find?
- Again, using the image of the fraction wall, how else could you write $\frac{3}{4}$?
- What other fractions do you know that are the same as $\frac{1}{2}$?
- Find some other fractions which are equivalent to $\frac{3}{4}$
- Can you find any "rules" for working out equivalent fractions?

Activities such as this encourage children to come to their own understandings of the rules that govern working with fractions. Many children will openly share their opinion that "fractions are hard." We wonder if this is because they have become used to simply being given rules they should follow rather than seeing for themselves that the rules actually make sense.

Ratio and proportion

Many pre-service teachers seem concerned when they are asked to teach ratio and proportion. This has always interested us as it is an area of mathematics that we all have intuitive understandings about. In the book *Street Mathematics and School Mathematics*, Terezinha Nunes and her colleagues showed that street children in Brazil had well-developed ideas and strategies to work with ratio and proportion when they worked in the street markets but that they could not transfer the methods they employed out of school to contexts that were introduced to them in school. For example, try to complete the following conversion table (this activity is most effective if you carry it out with some colleagues):

Cups of sugar	1	2	3	4	5	10	20	50	100
Cups of fruit	2.25								

When you have completed the table, talk to your colleagues about the strategies that you used. Many people find that using "doubles" is effective, so they will complete the 2 and 4 columns first. You can than find how much fruit is needed to make jam with 3 cups of sugar by adding 2 + 1 cups of fruit or multiplying by 3. If you know 3 and 2 you can find 5. Once you have 5, doubling can allow you to complete the table. Try this sort of strategy to complete this table:

Lemon	1	2	3	4	5	10	20	50	100
Kilograms of sugar	1.75								

It is important to use the language of ratio and proportion from the early stages so that when the ideas are introduced more formally, later in primary school, children have the linguistic base on which they can build their understanding when they are introduced formally to ratio and proportion. A formal learning objective may be phrased as: students should use the vocabulary of ratio and proportion to describe the relationship between two quantities (for example: "There are two red beads to every three blue beads, or two beads in every five beads are red").

We hope you can see how this vocabulary can be introduced at a day-to-day level across the school, perhaps at the beginning and the end of school days looking at things like school lunches and sandwiches; numbers of boys and girls or how many people who walk to school or come in the car.

The previous big ideas have introduced you to the mathematical content that you need to come to understand in order to teach young children to count; to support children in coming to understand place value; to introduce them to ideas of fractions, decimals and percentages and finally to make sense of ratio and proportion. The following teaching point offers you support in teaching these ideas in your classroom.

Teaching points

Teaching point 1: inconsistent counting

Jess asks Holly how many ducks there are in the pond:

Holly says it's easy and counts 1, 2, 3, 4, 5, 5, 7, 8, 9. Holly has not grasped either the one-to-one principle as she counts the big duck twice; perhaps in her mind she sees the size as important and so counts twice. She also misses 6 out of the count. This shows she does not yet realise that you always have to use the number names in order.

The best way to support students in overcoming this misconception is by giving them plenty of experience in counting. Try to use everyday objects that can be arranged in many possible ways for the students to count. Asking them to check each other's counting is a useful way of their supporting each other. Counting with the students is also important as you can model touching each object as you count it. Ask the students to count the same set of objects several times, arranging the objects differently each time. This will help them come to an understanding of the order irrelevance principle. It is also important to teach the students lots of counting rhymes and songs. This supports the development of the stable order principle.

TAKING IT FURTHER: **FROM THE RESEARCH**

In their paper, *Young Australian Indigenous students engagement with numeracy: Actions that assist to bridge the gap*, Elizabeth Warren and Eva DeVries argue that young Indigenous Australian students do not commence school with the same understanding of number concepts as non-Indigenous counterparts. This can mean that Indigenous students "fall behind" as the current mathematics curriculum takes a Westernised view of mathematics rather than an Indigenous view of mathematics, learning and teaching. The actions that they suggest include an initial focus on oral language in particular positional language, which allows all students to begin school on a similar footing. This also allows teachers to focus on developing mathematical language in a context familiar to Indigenous students from home.

They also suggest that mathematical language should be developed through discussions and experimentations in play-based contexts, which allows all students to begin to use formal mathematical language in a non-threatening environment.

Finally, exploring key aspects of number in numberless environments supported the development of counting and the language of counting. The fraction wall shown earlier in this chapter would be such an example.

Teaching point 2: miscounting on a number track

Sam is playing a game and he sometimes starts counting from the square that his counter is sitting on. This will cause a problem for him later when he uses a number line to support his mental calculations. This may be the reason that some children give incorrect answers for addition calculations. For example, they may write

$$5 + 3 = 7$$

1	2	3	4	5	6	7	8	9	10

This is because they counted 1, 2, 3, 4, 5 and then started on the 5 and count 5, 6, 7.

Again, the best way to support students through this is to use number lines regularly in many different contexts: frogs on lily pads; house numbers. It is particularly useful to ask young students to move on number tracks that you create in the classroom or the playground. Playing games is also important. The students very quickly correct each other if they are counting incorrectly in a competitive situation.

Teaching point 3: directed numbers

A **directed number** is one which has a plus or minus sign attached. This tells us whether it is a "positive" or "negative" number. So +7 means "positive" 7 and –3 means "negative" 3. A pre-service of Tony's was researching directed numbers before planning a lesson for her Year 6 class. She remembered being told by a teacher in her secondary school that, "two negatives make a positive." She had written –4 + –2 = +6 and said to him, "that can't be right."

Tony illustrated the calculation using a number line, moving to the right on a number line represents addition and to the left subtraction. So, I start with the arrow at –4. I am adding so I expect to move to the right; however, I am not adding 2, I am adding –2, so in this case the minus sign reverses the direction and the arrow moves to –6. So –4 + –2 = –6.

Primary age students should be using directed numbers to find the difference between positive and negative integers in context or to count down in whole number and decimal steps extending beyond zero. An integer is a number that has no decimal or fractional part. We sometimes call them whole numbers. An integer can be either positive or negative. Integers should not be confused with "natural" numbers. The **natural numbers** are all the positive integers, that is 0, 1, 2, 3, 4 and so on. Students may also be expected to position positive and negative numbers on a number line and to be able to use the < and > symbols to state inequalities. (An **inequality** is a statement showing which number is greater or less than another: < means "less than" and > "greater than." So –2 < +7 and –4 > –9.) This is helpful when students come to explore more complex equations at later stages of their learning.

The most appropriate way to support students with all of these ideas is to use a number line that extends beyond zero. We would suggest that it is appropriate that the number lines used for display or used to support children's calculations in the later stages of primary school, should always extend above and below zero by the same number. This develops the student's visual representation of the number system to include negative numbers. One useful context to use in support of the students' understanding of negative numbers is temperature.

Measuring temperature

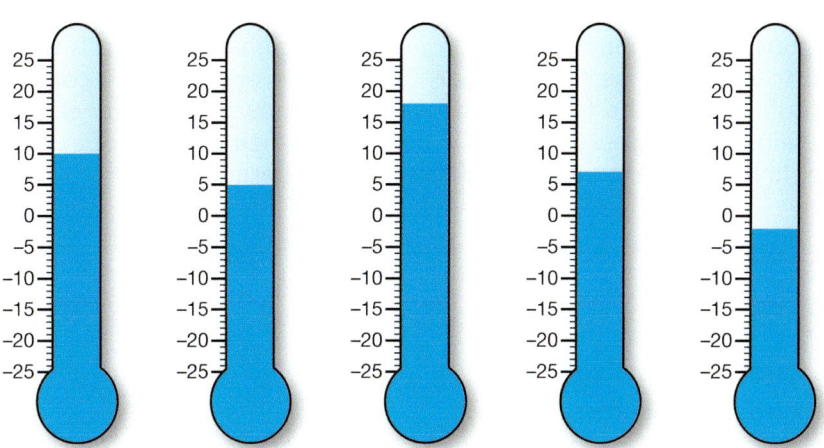

Jan	Feb	Mar	Apr	May	Jun	Jul	Aug	Sep	Oct	Nov	Dec
15	14	12	8	3	0	-2	-1	2	6	10	13

This table shows the average maximum temperature for each month at Mount Kosciuszko in NSW. Which months do the thermometers represent? Shade in the blank thermometers to show the average temperatures in other months.

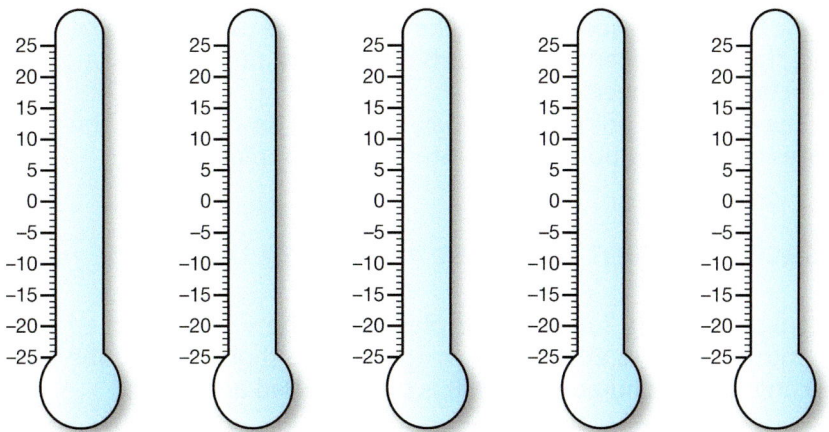

This example shows how directed numbers can be illustrated using temperature. It is even better if you use temperatures for the local area in which you teach. To provide a wider range of temperatures use other contexts that are meaningful for the students. If your school has a connection with a school in another area with a different climate this is ideal.

The thermometer provides a representation of a number line for the learner and the students are asked to draw their own number lines/thermometers to support them in developing their own use of number lines to model calculating using directed numbers. It is helpful to describe "getting warmer" as addition and "getting colder" as subtraction. The following activity shows how this activity can be developed to support students in developing their own use of a number line to come to understand directed numbers.

Properties of numbers and number sequences

You will need:

● cards with numbers 1–20
● cards with numbers 10–20
● two cards, one with a + and one with a -

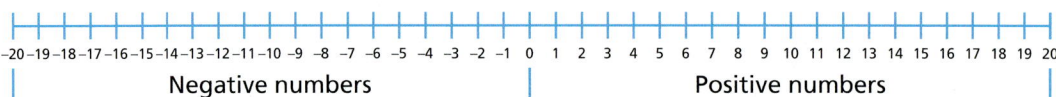

You will be creating number sequences with seven numbers.

● pick a card between 1 and 20 to get the size of the steps
● pick a card between 10 and 20 to get the starting number
● pick a + or – card to give you the direction of the sequence.

Draw the number sequences on the number line using arrows to show the "hops" between the numbers. Extend the number line in either direction if you need to.

The teaching points so far have focused on counting. The next series of teaching points move into the area of place value.

Teaching point 4: errors in writing numbers

Zeynab counts 19 lollies in the play shop. She writes down 91. Her friend Jasbir looks at the date written on the shopping list. She says, "I can read that number, 2008, it is two hundred and eight."

This is a common misconception and comes from children literally "reading" the number. Zeynab sees the "9" digit first and knows she should say nine–teen. So, she writes "9" and then adds the "1" to signal "teen." Similarly, Jasbir notices the "200" and then the "8" so she sees 200 and 8.

Counting activities that involve students grouping objects in tens and then finding the appropriate numeral on a number line will help these students. For example, you could give students 18, or 23 or 47 kidney beans, ask them to count them and then find the appropriate place on a number line and write on a card how many beans there are. As students get older, the use of place value cards to support them in partitioning numbers will help them understand the value of each digit in a number.

Teaching point 5: confusion about the use of 0 as a place holder

Some students become confused about the use of 0, particularly in decimals. For example, a student may think that 1.5 and 1.50 are different numbers. This can happen, in particular, when a student is using a calculator to work out money problems as often calculators will "remove" the 0 in $1.50 when it is entered. Many calculators will "remove" the 0 if you type in 1.50 when you press the operation key. It is worth doing this deliberately with students so that they notice it happening. Using a resource, such as arrow cards, can support your students in coming to an understanding of place value as they show the value of each digit in a number when you remove the arrow cards. The best form of the activity is to let your students have a set of arrow cards (you can download these from the companion website) and set each other challenges to make numbers. Once they have represented the number with the arrow cards, they should say the number aloud. It is also helpful to make a series of numbers and ask the students to order them in order of size.

Place value

You will need:

- a partner
- place value cards
- mini-whiteboards.

Read the following numbers to each other. Then make them with your place value cards. Read aloud the number from your cards to make sure it is the same as the one written in words.

The first one has been done for you.

1 Twenty-four thousand three hundred and sixty-one

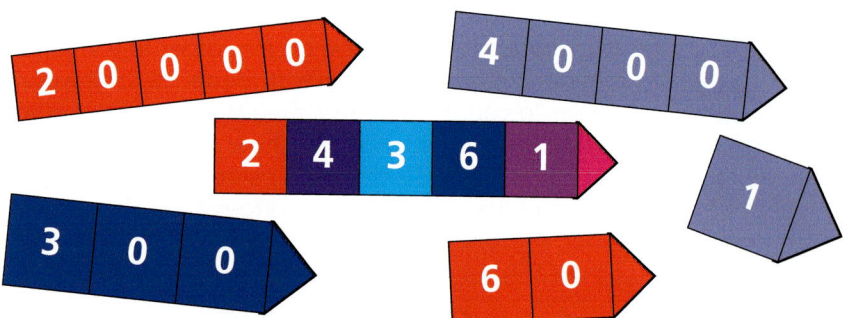

2 Eighteen thousand five hundred and nineteen
3 Thirty thousand seven hundred and eighty-two
4 Seventy-two thousand and forty-seven
5 Forty-five thousand eight hundred and twenty
6 Eleven thousand two hundred and fifteen
7 Fifty-six thousand and seven.

Teaching point 6: multiplication and division by powers of 10

One fairly common "rule" that some primary teachers share with students is the idea that when multiplying a number by ten, you just need to "add a zero." And while at first glance this may seem to be a helpful tip to share in the short term (e.g. 23 x 10, so if I just add a zero to the end, it will be 230), it will lead to students developing a misconception. The same rule that seemed to have served students so well when working with whole numbers quickly falls apart when moving to decimals. For example, consider trying to work out 1.7 x 10 using this "rule." Some students will think the answer is 1.70, which is clearly incorrect. Others may "add the zero" at end of the whole number and arrive at a solution of 10.7.

A focus on moving the digits one place to the left when multiplying by ten and moving to the right when dividing will overcome this misconception. Use place value mats or number sliders when carrying out activities that involve multiplying and dividing by multiples of ten.

Thousands	Hundreds	Tens	Ones

Thou-sands	Hundreds	Tens	Ones	Tenths	Hun-dredths	Thou-sandths	Ten Thou-sandths	Hundred thou-sandth	Mil-lionths

You can also model multiplication and division by multiples of ten using students and large number cards. The only rule is that the person holding the decimal point can never move. The following example shows how you can introduce the idea of moving digits and not adding a zero to your students.

Multiplying and dividing by 10

1 Write a single digit in the middle of a page

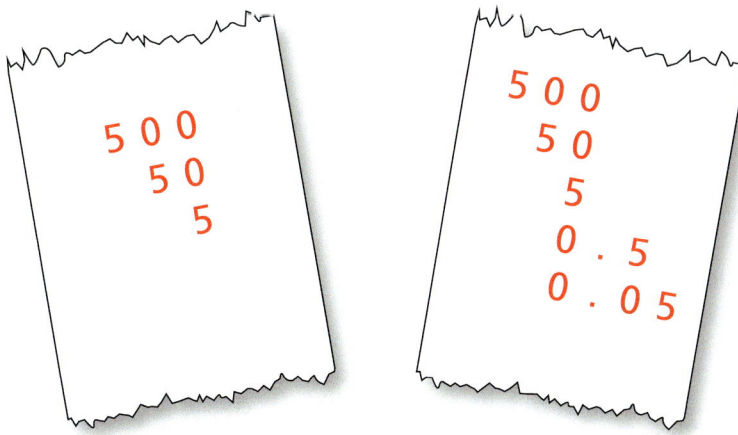

2   Multiply this number by ten and write this number above the first number. Repeat working up the page, until you have five numbers

3   a   Now divide your original number by ten. Write the new number below your original number

    b   Repeat, working down the page. What patterns can you see?

4   What happens when you divide a whole number with up to four digits by 10 and then multiply the answer by 100?

Teaching point 7: ordering decimals

Kev asks a group of students to order the following decimals:

1.7, 2.52, 1.65, 2.435, 0.812 One of the students writes:

1.7, 1.65, 2.52, 0.812, 2.435

The student has ignored the decimal point to order the numbers. So, the student sees seventeen for 1.7 and two thousand, four hundred and thirty-five for 2.435. If students have this difficulty give them a place value grid to place the numbers on. For example:

Ones	Decimal point	$\frac{1}{10}$	$\frac{1}{100}$	$\frac{1}{1000}$
0	.	8	1	2
1	.	6	5	0
1	.	7	0	0
2	.	4	3	5
2	.	5	2	0

Encourage the students to use the zeroes as place holders and look for the patterns. When they write the numbers in ascending order outside the grid, they can remove the zeroes that are not needed and discuss when we need to write a 0 and when it can be omitted.

Teaching point 8: rounding numbers

A student was asked to round 795 to the nearest 10. They had written 800 and crossed it out and written 790. When asked why they had crossed it out and they said it could not be right as they had been asked to the nearest 10 and not the nearest 100; they also knew that 810 was, "too many" so they had gone back to 790. Not recognising multiples of 100 as possibilities for multiples of 10 is an error that many students show.

A similar error with decimals would be a student who rounded 0.3 to 1.0 when asked to round to the nearest whole number. Here they do not see zero as a whole number and so ignore the convention of rounding up from 5. However, 0.3 to the nearest whole number is 0 as it is nearer 0 than 1.

An empty number line is a great resource to use when working with students on rounding numbers. A useful strategy is to ask students to set each other challenges. They can write numbers down for each other to place on the number line and then round to an accuracy either you decide or the students themselves decide. The following activity gives another example of an activity you could try. This can be adapted for rounding to any degree of accuracy.

ROUNDING

1. When I round to the nearest 10, I get an answer of 130. What are four possible numbers that I might have started with? Place these numbers on the following number line

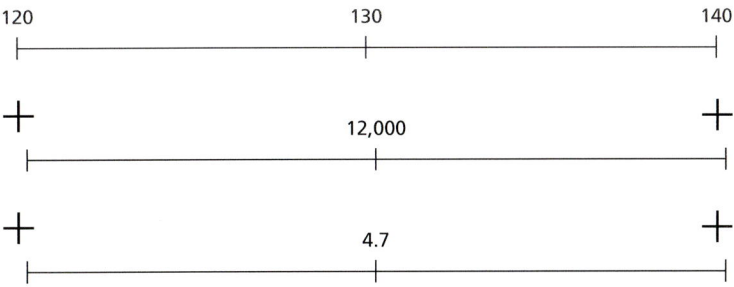

2. When I round to the nearest 1,000, I get an answer of 12,000. Write down five possible numbers I could have started with and place them on the number line

3. When I round to the nearest tenth, I get an answer of 4.7. Write down six possible numbers I could have started with and place them on the number line.

Teaching point 9: fraction names and writing fractions

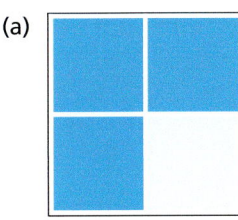

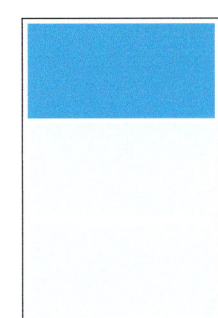

When faced with naming or writing the examples of the previous fractions, students may say many things. A response to naming the fraction represented by (a) might be "three-fourths." Here the student has remembered the convention for the denominator (number of parts in total) and the numerator (number of shaded parts) but has applied the convention of naming numbers that they are used to. Similarly, we have heard fraction (b) called "a threeth." It is important to listen carefully to the names that students give to fractions. In the previous examples the students had an understanding of the ways in which the names are formed and so have understood the underpinning idea; they simply need help in selecting the appropriate language. This can best be done through the use of classroom displays and the careful modelling of key words.

Sometimes, however, the naming and writing of fractions may point to a deeper misconception. Another student looking at fraction (b) wrote $\frac{1}{2}$. They also wrote $\frac{3}{1}$ and then crossed it out to write $\frac{1}{3}$ for fraction (a). Here they are noticing shaded and unshaded parts of the diagrams but are not seeing a fraction as a division of a whole. So, in shape (b) they saw one shaded part and two unshaded parts. They knew that $\frac{1}{2}$ is a common fraction and so wrote the fraction as shaded parts/unshaded parts. The same thinking led to their response for fraction (a). In this case, however, they saw the answer $\frac{3}{1}$, coming from shaded parts/unshaded parts but said that this, "didn't look right," so they inverted the fraction to a fraction that they had seen before.

An understanding of naming and writing fractions is key to successful calculation. It is probably worth repeating that a fraction is made up of:

Numerator	The number of parts of the "whole"
Denominator	The number of fractional parts the "whole" has been divided into

It is often helpful to remind students that the line separating the numerator from the denominator refers to division. So, in the previous examples, (b) can be seen as one out of three or one divided by/into three. Similarly, for (a) we can say, "three out of four."

Portfolio task 4.6

Write down all the errors that you can remember students making when they have named or written down fractions.

For each example decide whether the student's misconception was based on a misunderstanding of the language of naming and writing fractions or a lack of clarity around what defines the numerator and denominator of a fraction.

Add these notes to your portfolio.

Teaching point 10: fractions as equal areas

A pre-service teacher asked her students to draw as many examples of $\frac{1}{2}$ as they could by shading in a square. There were many imaginative and correct answers such as:

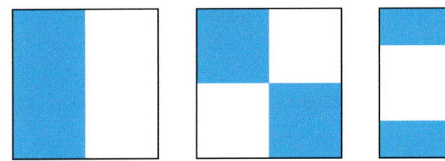 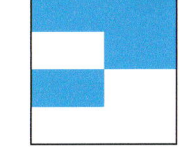

This shows that both the pre-service teacher and the young student understood that $\frac{1}{2}$ could be represented by any 4 parts out of the 8 that made up the whole; $\frac{1}{2}$ does not have to be symmetrical or obey a particular pattern.

Other students had used diagonal lines, which allowed them to show many more views of $\frac{1}{2}$. However, one boy had drawn the following:

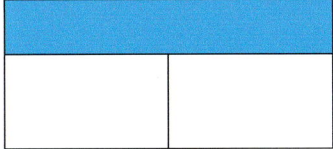

At first, he was adamant that this was $\frac{1}{2}$ as there were, "two out of four shaded in." He was persuaded that this was not a half when asked if he would be happy to have the "smaller half" if this was a bar of chocolate.

This misconception often arises as informally we may call any sharing into two "halving." If we are splitting a cake, for example, we will ask our friend, "Do you want half?" without thinking the two parts have to be exactly the same size. So, when introducing fractions to students it is important to focus on the importance of equal areas.

The following activity focuses on this idea as well as making the link between fractions and decimals. This activity can only be carried out practically as the students will take great care to make sure things are fair.

CHOCOLATE FRACTIONS

Place three chairs at the front of the class.

Place one bar of chocolate on one chair, two bars on the next and three bars on the next.

Ask ten students to take it in turns to stand behind one chair making their choice of chair one after the other.

When all the students are standing behind the chairs, they can divide the chocolate on the chair equally between the children at that chair.

This activity leads to a great deal of discussion about fractions as equal areas. It can even make quite a good party game.

Teaching point 11: fractions must be less than one

Before exploring this teaching point in more detail try this task.

Portfolio task 4.7

Can you write down 2.75 as a fraction and a percentage?

Can you do the same for 1.5 and 4.9?

When students are asked to carry out this activity some find it difficult. Although they are used to seeing decimals greater than 1, they feel different about **improper fractions** (an improper fraction has a numerator larger than or equal to the denominator such as $\frac{7}{4}$ or $\frac{4}{3}$) or percentages larger than 100%. This is probably because we are so used to thinking of fractions as simply a part of a single whole rather than as a way of representing any part of a number between two whole numbers. So, we can write 2.75 as $\frac{11}{4}$ or even 275%. It is important to use fractional number lines with students and it is crucial that these number lines extend beyond 1. This allows you to represent fractions and decimals together, so students can see how they can move between decimals and fractions in order to carry out comparisons and calculations.

Teaching point 12: ordering fractions and decimals

Students often have difficulty ordering fractions. For example, they may think that $\frac{2}{5}$ is larger than $\frac{2}{3}$ as they notice that 5 is larger than 3 and they do not have mental images of the two fractions to fall back on. An effective method to support students in ordering fractions and decimals is to use an approach that uses multiple representations of fractions and decimals. This image shows you how you can represent fractions and decimals as areas and on a number line.

The following activity will help you to support students in coming to an understanding of fractions, their decimal equivalents and how

to order them. Here, they are encouraged to draw on their mental images of fractions and decimals to place them on an empty number line and draw on their current knowledge of fraction/decimal equivalents to calculate new equivalents. It would be even better to carry out this activity practically using a physical washing line in your classroom.

FRACTIONS AND DECIMALS

1   Look carefully at the pegs on the washing line. Estimate which fractions they could represent. Copy the numbers onto a large sheet of paper. Label the points A to H with their fractions and equivalent decimals

2   Working with your partner, look at the gap between peg B and peg C. Think of a fraction that will hang on the line in the gap. Predict where it will hang. Calculate the decimal fraction to check it will fit before you hang it on the line

3   Find a fraction to hang in each gap between the pairs of letters. Remember to calculate the decimal equivalent before you add it to your line

4   Look at the gaps between the pegs. Now find a space to add three more fractions that have a different denominator to any that are already on the line.

Teaching point 13: the meaning of percentage

Tony began his career teaching in secondary schools and still remembers the following comment from one of the fifteen-year-old boys he taught. He had arrived at the mathematics lesson fresh from a French test, he was beaming and said, "We've just got the results of our French exam. I did really well, the teachers said. I got 85%. The only trouble is that I don't know how many it was out of." Like fractions, students have intuitive understandings of percentages. This understanding seems to have improved in recent years, with students having increased exposure to a range of devise, where their battery life is shown as a percentage. More specifically they will have seen 50% and may know that this is equivalent to a half. They may well have come across 10% in the context of shopping and prices that have been reduced. A good starting point for working with fractions is to explore student's current understandings of percentages. Look through the papers and magazines that they read. Ask them to bring in examples of percentages so that you can set a real context for the exploration of percentages. By the time that percentages are introduced they should have an understanding of the equivalence between fractions and decimals, so it is important immediately to make this link as well. Of course, a percentage is always out of 100, with *per* meaning "for every" and *cent* "one hundred."

Another example of the misconceptions that can arise with percentages is confusing a percentage "of" (for example, 20% of 60 = 12; 20% is the same as $\frac{2}{10}$ so we can write $\frac{2}{10} \times 60$, a calculation I would carry out mentally: $\frac{1}{10}$ of 60 is 6 so $\frac{2}{10}$ is 12) with the idea of "out of." For example, if I get 15/20 in a test, I have achieved a mark of 75% ($\frac{15}{20}$ can be cancelled down to $\frac{3}{4}$ which is equivalent to 75%). A good friend once telephoned Tony. He was a police officer and was having to submit a report on crime figures. He said that he knew the old figures for burglary and knew the increase but, "Couldn't remember which one to divide by 100." This is a good example of trying to remember a rule or a procedure that we were taught at school but being unable to see how this applies to a real context.

Teaching point 14: recurring decimals

As your students use calculators to carry out conversions between fractions and decimals or to carry out complex calculations, they will discover recurring decimals. The most common recurring decimals are thirds, sixths and ninths. When we write recurring decimals, we write enough digits after the decimal point to show the recurring pattern and insert a dot on the last decimal place to show it is a recurring pattern. For example,

1.3333 would be written as $1.\dot{3}$

or

1.214214214 would be written $1.\dot{2}1\dot{4}$

Teaching point 15: confusing ratio and proportion

A pre-service teacher asked her students, "What proportion of this class are boys?" In her class of 25, only 10 were boys. One of the students replied, $\dfrac{10}{15}$ (ten-fifteenths). This response showed some understanding of the idea of ratio and proportion. The student knew the answer should contain the numbers of boys and girls, and knew there was a link to fractions, but had confused ratio and proportion. Many adults find the difference between ratio and proportion difficult to remember. The definitions are given below.

> **Ratio:** a ratio compares part to part. We use a phrase like "to every" or "for every". So, in the previous example there are 10 boys to 15 girls, so the ratio is 10:15.

As with fractions we can cancel this down by dividing each number by five. This gives us a ratio of 2:3. We might also say that for every two boys there are three girls.

> **Proportion:** proportion compares a part to the whole. In the previous example there are 10 boys in a class of 25, so $\dfrac{10}{25}$ or, by cancelling, $\dfrac{2}{5}$ of the class are boys. We would say that 2 out of 5 of the people on the trip were boys.

Similarly, 15 out of 25 are girls, which is the same as 3 out of 5.

Teaching point 16: adding rather than multiplying when increasing ratios and proportions

Another error you may come across when teaching ratio and proportion is students reverting to adding rather than multiplying when increasing ratios and proportions. For example, when given a recipe for soup that contains 2 onions in a recipe for 3 people and asked to work out how many onions would be needed for a recipe for 12, they add 9 onions, giving an answer of 11. The answer is, of course, 8 onions. We are cooking for 12 people so we must multiply 3 by 4 to get to 12 as the original recipe was for 3 people. So, we need 4 times as many onions, giving us 4 × 2 = 8 onions.

Carrying out practical cooking activities is a great way to "practise" calculations involving ratio and proportion.

Portfolio task 4.9

Look back over the misconceptions you have explored in this chapter.

Which of the misconceptions did you share?

How do you think you developed this misconception?

What might a teacher have done to support you in overcoming the misconception?

Before moving on to the "In practice" section, try to write a convincing explanation to support a colleague in understanding why this misconception is incorrect.

Add this to your portfolio.

In practice: build it

The lesson plan which follows was used to support a group of students early in their understanding of place value in a Year 1 and Year 2 multi-age classroom. Following the plan is an evaluation of the lesson that explores how successful the plan was in supporting the students to develop their knowledge, skills and understanding.

Learning Intentions

To recognise the value of a digit based on its location within a number

To construct a two-digit number using materials that supports place value understanding

To count a set of objects and identify the number that represents the set

To use reasoning to explain efficient counting methods, strategies or materials

Key Vocabulary

digits, tens, ones, groups of, equal to, lesser than, more than, compare, skip counting

Resources

Number cards 21–49

Materials that can be used for counting (Unifix cubes, counters, pebbles, teddies)

A collection of specific base-ten manipulatives (MABs, icy pole bundles)

Context for lesson

This lesson was carried out early in a unit that focused on connecting additive thinking and counting with place value. It was used to as a means to discuss the value of the digits in two-digit numbers and to reflect on the efficiency of counting in groups of tens and ones. It was noted prior to conducting this lesson that when faced with a larger, unfamiliar numbers, students tended to revert back to counting by ones as they deemed it more reliable when, in fact, this practice often led to more calculation errors.

Warm up activity: double hat-trick

In this game, two students face head-to-head in front of the class to try and place four of their numbers in a row on a 0–100 number line without their opponent disrupting their sequence.

Dat rolls the two 10-sided dice which lands on 4 and 8. He needs to decide whether to construct a 48 or an 84 and place one of these numbers on the number line. This is a great opportunity to clarify the vocabulary of "digit" and "number." Dat chooses to place 48 in the middle of the number line. An important point to note here is to focus the students' attention to using estimation and benchmark numbers to locate their numbers on the line with precision. His opponent Olivia rolls a three and a one. She decides to place the number 13 on the line in a different colour. Dat takes his next turn and rolls a two and a four. He chooses to make 42 and, when asked why, he enthusiastically reasons that the chances of Olivia squeezing her way between his 42 and 48 were slim. These anticipated, posed questions highlight the strategy and forward-thinking involved, turning this fun game into a rich, engaging maths warm-up.

The game continues back and forth between the two players until one of them has secured four consecutive, uninterrupted numbers on the number line.

Launch: build it

I placed the numbers 21–49 in a bag, leaving out multiples of 10. This careful selection of relatively small numbers ensures that students will not be distracted by the task of finding enough objects to build their secret number, whilst also allowing an accessible entry point for all students.

Students then took a "lucky dip" inside the bag and pulled out and held onto that number, keeping its value a secret.

Explore: build it

The class was then directed to use objects from around the room to make their secret number, with an emphasis on using materials that would support efficient counting (without explicitly naming them e.g. Unifix cubes, icy pole bundles, MABs).

After they had all made their secret number on their table, they attached their own name to their representation as a means of identifying the builder. Students then walked around the room and worked out each person's secret number by counting the number of objects in their identified spots, recording it on a class list of names. Whilst students were walking around the room, I prompted them with questions to think about which we would later reflect on, such as "Whose representation was easiest to count and why?"

As the students were carrying on with their task, I walked around and took photos of specific representations that used different materials and were organised in different ways to support the whole-class reflection that was to follow.

Summary: build it

After it was clear that the students had enough time to count and record the secret numbers of a reasonable number of classmates, we gathered together to reflect on the session. The conversation was guided by the following general questions: "What materials did you choose to build your number and why?" "How did you choose to organise the parts of your number?" leading to more reflective questions: "How did the organisation of materials on the table make it easier or harder to count?" and more specific place value questions such as "So what does the digit 3 in the number 37 represent?" "How does knowing this make it easier to count?"

Rationale and evaluation: build It

The open-ended nature of this task in terms of what materials they could use and the way in which they were going to display their number made it engaging for all students. Some students needed reminding that the emphasis was on building their number so that it was easy for others to work out as they were preoccupied by the idea of it being a "secret" number initially.

It was interesting to note that, whilst gathering their objects, most students counted their own collection by ones but, when arranging it for others to count, chose to organise it more methodically, though this wasn't always by place value parts. Many students assembled their objects into discrete groups of twos, some by threes (though they had trouble counting by threes to find the total). Some arranged their materials into array-type representations. Some students who chose the icy pole bundles and base-ten blocks knew and used the concept of decomposing into tens and ones, whilst others tried using each icy pole bundle of ten, for example, as a representation of one which is a common misconception. This made for a great talking point during the reflection.

Displaying the photos of the representations on a large screen made the discussion visually engaging and conceptually richer as a result. It was easy for students to compare the different representations and use reasoning to explain why some were easier to count than others.

Audit task

Devise a lesson plan that is appropriate for a group of students you are working with. The focus should be "counting and understanding number." Construct a lesson plan using the proforma on the companion website. Teach the lesson and then evaluate it carefully with a focus on student's learning and misconceptions. If you can, use ICT to support the learning, but only do this if the ICT enhances the learning. Give evidence for the effectiveness of the ICT in the evaluation. Add this lesson plan and evaluation to your subject knowledge portfolio.

Assessing counting and understanding number

For early learners it is most appropriate to assess their understanding of counting and understanding number through practical activity. For example, you can give 20 students in your class cards each with a separate number between 1 and 20 on it. Pick students in the class to give them instructions so that they are sorted into the correct order. Similarly, you could ask students to stick numbers onto a picture of a street with house numbers if you want to include odd and even numbers, or you can use a block of flats if you simply want to order whole numbers. Ask the students to stick the numbers on the doors so that the mail carrier can find the correct houses to deliver to.

Estimation games are great for assessing students' understanding of number and place value. Fairground games such as guess the number of lollies in a jar can be used to assess the students' sense of number. You can also observe how they count the lollies. Do they count in twos, fives or tens? They can then round to the nearest ten.

Real-life data such as heights of tallest buildings, lengths of rivers or sizes of populations can be used to explore larger numbers and assess students' skills of rounding. Similarly, use very small measurements of micro-organisms to round to a given number of decimal places.

Matching equivalent representations of fractions, decimals and percentages is an excellent assessment activity too. To assess students' understanding of finding equivalents, leave some of the equivalence cards blank so that the students have to fill them in for themselves. It is important to ask students to order these numbers as a part of this activity so that they come to see how useful it is to move between different representations. Many students will find it difficult to order fractions but much easier to order decimals or percentages. Having this understanding can also help if your students still have difficulty in ordering fractions by looking at the numerator and denominator.

Cross-curricular teaching of counting and understanding number

Almost any planning for a class or school event will involve the students in counting and applying their understanding of number. Students of any age can be involved in planning and organising events in the classroom or for the wider school community.

If you are working in an early years setting there will be many occasions when you wish to invite parents and carers to events in the classroom. The students can be involved in making sure there are enough seats, enough hangers for their parents' coats and, if you are serving food, making sure that there are enough places set so that everyone has a knife, fork and plate. You can even involve the students in making refreshments. Making buns with a cherry on top leads to lots of counting activities – particularly if you cut each cherry in half to make the buns look even better.

Older students can get more involved in making the refreshments. You can use recipes for smaller numbers of people and use ratio and proportion to ensure that you have enough ingredients to make sure everyone has enough to eat. If you are buying pizzas the students will need to think about what fraction of each pizza everyone will want and then use this to calculate how many pizzas to buy.

You may want to divide your class up into separate groups each with different responsibilities. One group can organise the invitations and make sure the class knows how many guests to expect. Another group can be in charge of refreshments, both buying and providing sufficient food and drinks for all the guests. A third group will need to organise hospitality during the event and make sure there is enough space for the event to take place. Finally, one group will need to be in charge of the budget for the event.

Summary

The aim of this chapter was twofold: to support you in understanding how students learn to count (something it is hard to recall from our own experience) and to offer a range of activities that will support you in teaching young learners this most basic of skills. We would also hope that these activities will allow you to observe young students coming to an understanding of counting. The big ideas of "counting," "place value," "fractions, decimals and percentages" and "ratio and proportion" have also been explored with teaching points to accompany each of these big ideas. We hope that by exploring each of them holistically you are able to make the connections between and within them. That is, you can see how fractions, decimals and percentages are all simply different ways of writing numbers and that choosing the most appropriate representation can make calculating simpler.

Reflections on this chapter

We opened the chapter by suggesting that counting and understanding number is an area you may find difficult to teach as it is something we cannot remember learning ourselves. If we had to pick a key idea from this chapter, it would be the five principles underpinning counting. These principles are observable when we watch students beginning their journey to counting confidently and support us in deciding how to structure the learning experience. The more you observe students at these early stages, the more you will be able to see these principles in practice. This area is also one in which we may bring our own misconceptions from our own experiences.

How many of us were told to "Add a zero" when multiplying by ten, or that "Two minuses make a plus" when we were introduced to directed numbers? Our teachers who used these stock phrases were showing that they were not confident in their own mathematical understanding. We would hope that the teaching points within the chapter have offered you alternative ways to describe these processes, explanations that will not lead to misconceptions.

Going further

The following books and research papers have been referred to in this chapter. Here is a brief overview in case you want to explore these areas in more detail. We have also included the full reference for each book.

Children and number: difficulties in learning mathematics

MARTIN HUGHES

Reading this will give you a new perspective on young children leaning number, in particular the skills of counting and understanding number. Martin Hughes shows how young learners bring a wealth of knowledge about counting and understanding number with them when they start school. He argues that it is important for teachers to support students in making links between their own informal understandings and the more formal language of school mathematics. He includes lots of ideas for activities that teachers can draw on to make this possible.

Hughes, M. (1986) *Children and Number: Difficulties in Learning Mathematics.* Oxford: Blackwell Publishers.

Mathematics with reason: the emergent approach to primary maths

EDITED BY SUE ATKINSON

This is a very practical book that aims to support teachers in connecting their teaching with the ways that students learn mathematics outside the classroom. As with the Martin Hughes book, it takes the approach that students develop understandings of number at a very early age and that it is the teacher's role to draw these understandings out and to develop them. There are contributions from a wide range of teachers and other mathematics educators who describe how they put these ideas into practice in their classrooms.

Atkinson, S. (ed.) (1992) *Mathematics with Reason: The Emergent Approach to Primary Maths.* Abingdon: Hodder and Stoughton.

NRich

The Nrich website describes itself as the home of rich mathematics. It is part of the University of Cambridge. The website describes rich mathematics as mathematics which offers opportunities for learners to be creative and curious through:

- exploring and noticing
- asking their own questions and conjecturing
- visualising and representing
- reasoning, justifying and proving.

The activity we shared with you earlier in the chapter can be found at https://nrich.maths.org/4519 Accessed 9 December 2019.

Teaching and learning early number

IAN THOMPSON

This has been described as a radical and influential book by the *Times Educational Supplement*. Although solidly based in research, it is very accessible and offers lots of practical ideas that teachers of young children can use to support young learners in the early stages of learning about number and counting. It includes chapters exploring learning and teaching number through play and assessing students' knowledge of number skills through interviews.

Thompson, I. (2008) *Teaching and Learning Early Number*. Maidenhead: Open University Press.

Exploring the use of mathematical manipulative materials: is it what we think it is?

LINDA MARSHALL AND PAUL SWANN

Linda Marshall and Paul Swann, from Edith Cowan University, show that teachers' subject knowledge is vital in ensuring that manipulatives are used effectively to support students' developing understanding of place value.

Marshall, L. and Swann, P. (2008) *Exploring the Use of Mathematics Manipulative Materials: Is It What We Think It Is?* EDU-COM 2008 Conference. Available at https://ro.ecu.edu.au/cgi/viewcontent.cgi?article=1032&context=ceducom Accessed 24 May 2021.

Young Australian indigenous students engagement with numeracy: actions that assist to bridge the gap

ELIZABETH WARREN AND EVA DEVRIES

An exploration of how teachers can adapt their classrooms and their approach to teaching numbers to ensure that young learners from Indigenous backgrounds can be supported in their learning of a Westernised approach to counting and numbers.

Warren, E. and Devries, E. (2009) 'Young Australian indigenous students engagement with numeracy: Actions that assist to bridge the gap', *Australian Journal of Education*, 53(2), 159–175.

Self-audit

1   This question allows you to develop your use of number lines to explore ideas of rounding and estimation. Each of these numbers is the answer to a question asking you to round a number. You

should draw a number line to show the range of possible numbers you could have started with. For example, 2,300 has been rounded to the nearest 100 so the smallest number I could have started with would be 2,250 (we always round up from 5) and the largest number I could have started with could be 2,349 (as I would round 2,350 up to 2,400):

2,250 _____

2,300 _____ 2,349

- **a** 570 has been rounded to the nearest 10
- **b** 3,000 has been rounded to the nearest 100
- **c** 7.8 has been rounded to the nearest tenth
- **d** 10.75 has been rounded to the nearest hundredth
- **e** 0 has been rounded to the nearest whole number.

2 These questions explore your skills in expressing numbers as decimals, percentages and fractions. It is important that you feel comfortable moving among these three representations of numbers. Children who can move easily among fractions, decimals and percentages are better able to carry out complex calculations and have a clearer sense of number.

a

20%	$\frac{5}{8}$	0.25
$\frac{9}{10}$	2.8	$\frac{2}{3}$
$\frac{5}{9}$	75%	$\frac{1}{3}$
30%	$\frac{1}{2}$	0.7

Use the fractions, percentages and decimals to make questions with the following answers (e.g. 75% of 12 is less than 10 and 0.7 × 20 is between 10 and 25):

i	< 10	**vi**	= 15
ii	> 25	**vii**	= 0
iii	between 10 and 25	**viii**	= 0.5
iv	between 5 and 6	**ix**	= 20%
v	< 1	**x**	= 4/5

b Draw this number line:

Add the following numbers to the appropriate place on the number line:

$$\frac{5}{8}, 0.9, \frac{1}{3}, 0.66, 75\%, 65\%, \frac{8}{9}, 15\%$$

The next question introduces proportionality. This is the final key idea within counting and understanding number.

3 Look at the following information about a group of pre-service teachers in a teacher- training course:

There are 25 students in a teacher-training course:

20 are female

10 are from minority ethnic groups

8 are living with their parents

4 are mature students.

I can write the following statements from this information:

- four out of five of the students are female.
- the ratio of students from minority ethnic groups is 2:3.
- 32% of students are living with their parents.

Write ten more statements using this information. Use fractions, decimals, ratios and proportions.

CHAPTER 5

NUMBER
Knowing and using number facts

Introduction

Mathematics is about much more than knowing facts. However, there are some facts that are useful as they support us in carrying out calculations quickly and in being able to check that our answers to problems are accurate. You may have played "Guess my number" with classes that you have taught. This is when you think of a number and the students in your class have to guess the number by asking questions. You can only answer "Yes" or "No" to their questions. The following exchanges probably ring true too:

Jess: I've written my number down on this piece of paper. Tom, you start asking the questions.

Tom: Is it 57?

Jess: It might take quite a long time if you just guess numbers. Michelle, have you got a question?

Michelle: Is it even?

Jess: Well done, that's a great question. No, it's not even. Megan, you next.

Megan: Is it odd?

Just as with the games where we ask people to guess famous people, success relies on good subject knowledge, here knowledge of number facts. It also relies on asking effective questions. The students in this class got much better at questioning, very quickly. After each game, they talked about what the best questions were, that is, which questions narrowed down the choices best. The students worked in pairs to come up with good questions before starting the next game. As an extension they tried to use the minimum number of questions possible. This class used to pick a "number of the day", and during the day, in a spare moment, students wrote down any number facts they could about this number.

DOI: 10.4324/9781003315155-5

The previous story is an illustration of knowing number facts, and, of course, we need to know number facts before we can use them. One of the most useful ways we can use number facts is to derive new facts. This is sometimes referred to as having good "number sense" (see the following box). When we think about a number, we get a sense of that number. So, we might look at 36 and see a number that has a lot of **factors**; a number that is a **square number**; a number that is divisible by 12 and so on. If we see numbers in this way, we are better able to work with the numbers, calculation becomes more straightforward and estimation comes easily.

So, this chapter emphasises how important it is to be able to use the facts that we can remember to derive new facts. For example, if I can remember that 2 × 6 = 12 I can work out that 4 × 6 = 24 and 8 × 6 = 48 by doubling and that 9 × 6 = 54 by adding another 6 and so on.

We want to emphasise that your students do not have to memorise a huge range of facts whilst they are being taught by you. Rather give your students time to make sense of the number facts they have been previously introduced and work with them to use the facts they already know to derive new sets of facts. Remember this when we discuss cognitive load theory later in this chapter.

TAKING IT FURTHER: **FROM THE RESEARCH**

The term "number sense" was introduced by McIntosh, Reys and Reys in 1992, in their article "A proposed framework for examining basic number sense". They explored the link between informal methods and the formal calculation procedures. The idea of number sense has been discussed widely and is an important area for further exploration if you wish to develop your understanding of how children draw on their own informal methods to come to an understanding of formal written methods. The most recent in-depth research around number sense can be found on Professor Jo Boaler's website. We would recommend viewing the videos on her website at www.youcubed.org/resources/what-is-number-sense/. This is an open access website on which you can find lots of activities to encourage the development of number sense in your students.

There is an ongoing debate about how much children should commit to memory. Jo Boaler has written a paper titled "Fluency without fear" (full reference at the end of the chapter). This revolutionary paper debunks the myths that we have to be able to have all our number facts at the tips of our fingers in order to succeed in mathematics. We imagine that many of you reading this book will not feel as though you can instantly recall all the facts you wish you could. In the paper Jo writes,

> *Some students are not as good at memorising math facts as others. That is something to be celebrated, it is part of the wonderful diversity of life and people. Imagine how dull and uninspiring it would be if teachers gave tests of math facts and everyone answered them in the same way and at the same speed as though they were all robots. In a recent brain study, scientists examined students' brains as they were taught to memorise math facts. They saw that some students memorised them much more easily than others. This will be no surprise to readers and many of us would probably assume that those who memorised better were higher achieving or more intelligent students. But the researchers found that the students who memorised more easily were not higher achieving, they did not have what the researchers described as more "math ability," nor did they have higher IQ scores.*

So, a key to knowing and using number facts is to develop the type of number sense described earlier in this chapter, in and with your learners. Developing this ease with numbers is what knowing and using number facts is about. Sometimes students seem to treat numbers with suspicion, as if they cannot be trusted. This happens if mathematics is presented to them as a series of tricks and rules to be learnt. If they can be presented with mathematics as inherently logical, they may begin to trust numbers, to see that they always behave in the same way. This chapter gives a wide range of examples of how mathematics is logical and how you can convince yourself and your students that it is logical.

The next section shows the development of student's knowledge of number facts as they move through primary school. It also shows you what you might expect the students in your class to have experienced before they meet you. These are the number facts that you can build on, the facts you can draw on as you encourage your students to deepen this knowledge. Of course, the easiest way to see what facts they bring with them is to have conversations with your students that allow them to demonstrate what they already know before you move on to teaching them new facts.

Progression in knowing and using number facts

Foundations for knowing and using number facts

At the early stages you should work with young children to help them observe relationships and patterns in numbers they see in their every-day environment, like house numbers, the numbers in blocks of flats, numbers on classroom doors, numbers in nursery rhymes and so on. In particular, you should work with them on building part-part-whole knowledge for each of the numbers up to ten. In addition to this, through using number lines, they will begin to see how the process for finding one more or less remains the same whatever number you start with. This is the beginning of seeing mathematics as logical and consistent.

Beginning to understand knowing and using number facts

At this stage your students will be skip counting in twos, fives and tens, starting at zero. They will use this skill to investigate num-ber sequences. They will know that you can partition numbers into hundreds, tens and ones and use this knowledge to count more efficiently

Becoming confident in knowing and using number facts

The next step is to learn how to derive and recall addition and sub-traction facts up to 10, all pairs with totals up to 20 and pairs of multiples of 10 up to 100. So, students will notice the patterns when you add

$$20 + 20 = 40 \text{ or } 50 + 50 = 100$$

and make the links with this pattern and the patterns of pairs up to 10 ($2 + 2 = 4$; $5 + 5 = 10$). Students will also begin to understand the difference between odd and even numbers and investigate their properties.

At this stage students will be making the link between halving and doubling and remembering multiplication facts for two, five and ten. They will also be beginning to use this knowledge to estimate and check their answers are sensible.

Once students have developed their ability to recall multiplication and division facts for 2, 3, 4, 5, 6 and 10 (you will see that this is different from simply being able to recite their times-tables) you can work with them so that they know all of the multiplication facts up to 10 x 10. Students at this stage should also recognise the patterns that allow them to recognise multiples of 2, 5 or 10 up to 1,000.

Extending learners' knowledge in knowing and using number facts

By the end of their time in primary school students will be able to recall or figure out multiplication facts efficiently and draw on this to multiply pairs of multiples of 10 and 100 (7 × 50 = 350 and 70 × 50 = 3500 and so on). They will be able to identify **factors** of two-digit whole numbers and use this knowledge to find **common multiples**. Factors of a number are the numbers that divide into that number. So, the factors of 12 are 1, 2, 3, 4, 6 and 12. A common multiple is a multiple that is shared by two or more numbers. So, a common multiple of 3 and 6 is 12 as 3 and 6 are both factors of 12. A common multiple is a multiple of both of the numbers.

Building on the understanding of decimals students will use their knowledge of multiplication facts to derive related number facts linked to decimals. (So, if I know that if 3 × 9 = 27 I can derive the following facts: 0.3 × 9 = 2.7, 2.7 ÷ 9 = 0.3.)

Students will also be able to derive **square numbers** (a square number is the result of multiplying a number by itself, so 9 is a square number because it is 3 × 3 and 25 is a square number because it is 5 × 5) and **triangle numbers**. They will also be able to recognise **prime numbers** less than 100. A prime number is a number with only two factors, itself and 1. So 13 is a prime number because its only factors are 1 and 13. The first 10 prime numbers are 2, 3, 5, 7, 11, 13, 17, 19, 23; "1" is not a prime number as it only has one factor. Any number that is not a prime number is described as a **composite number**.

The aim of this section is to support you in seeing how a student's knowledge of number facts gradually builds over seven or more years and how it forms a coherent and consistent landscape in which numbers behave as we expect them to. We can use this consistency and our current knowledge to derive new knowledge.

Big ideas

The two big ideas in this section are **Patterns** and **Rules**. Earlier in the chapter we discussed the development of number sense. This sense of number develops with an understanding of the patterns that exist within the number system and the rules which our number system obeys. Noticing patterns allows us to derive new facts quickly from those which we know, applying rules appropriately allows us to use the facts accurately. This skill is vital when we begin to develop our algebraic thinking. The word **algebra** is derived from the Arabic *al-jebr*, which means, literally, putting back together broken parts. Being able to notice patterns and seeing individual numbers in terms of their properties supports us when we need to create and solve equations.

There is a story of two famous mathematicians G.H. Hardy and Srinivasa Ramanujan at Cambridge University in England. Whenever they met, they would pick numbers and see who could find the most interesting facts about them. On one occasion Hardy had travelled to meet Ramanujan in hospital in a taxi which had the number 1729. Hardy picks up the story:

> I remember once going to see him when he was ill at Putney. I had ridden in taxi-cab number 1729 and remarked that the number seemed to me rather a dull one and that I hoped it was not an unfavourable omen. "No," he replied, "it is a very interesting number; it is the smallest number expressible as the sum of two cubes in two different ways."

We should not expect to know everything there is to know about every number, but being able to see numbers in many different ways allows us to manipulate them and to explore and understand pattern.

Patterns

Number lines and hundreds charts are useful resources to support students in noticing pattern. One of the first patterns that students may notice is what happens when we add or subtract one from a number. They can see on a 100 square and a number line that 1 more or 1 less can be represented by a movement of 1 digit to the right to add 1 and 1 to the left to subtract.

This image will support them as they explore more complex ideas, but they have already learnt an essential image: we add and subtract by moving forwards and backwards along a number line. This is the beginning of algebraic understanding. If I can see this pattern, I will be able to solve the equation

$$x + 1 = 9$$

For students in the primary school, we would substitute a cloud or a box for x, but the question remains the same, "If I add one to a number and get the answer nine, what is that number?" The following portfolio task asks you to notice the way that patterns on a hundreds chart can help learners carry out calculations. This discussion will be developed further in Chapter 7, which explores learning and teaching algebra in depth.

Portfolio task 5.1

Describe the following operations on a hundreds chart:

- adding 10
- subtracting 11
- adding 27.

Describe the move you make. For example, to add 11, "I move down one row and one column to the right." Add these notes to your personal portfolio.

1	2	3	4	5	6	7	8	9	10
11	12	13	14	15	16	17	18	19	20
21	22	23	24	25	26	27	28	29	30
31	32	33	34	35	36	37	38	39	40
41	42	43	44	45	46	47	48	49	50
51	52	53	54	55	56	57	58	59	60
61	62	63	64	65	66	67	68	69	70
71	72	73	74	75	76	77	78	79	80
81	82	83	84	85	86	87	88	89	90
91	92	93	94	95	96	97	98	99	100

The patterns that we see are all linked to the idea of place value, which was explored in the previous chapter, and a good understanding of place value is key in helping learners see the underlying reasons for the patterns that we notice. In the previous hundreds chart, all multiples of ten end in a 0, and multiples of five end in five or zero. Whenever you are introducing new ideas linked to number facts, draw on images of the numbers system such as number lines and hundreds charts to help children develop mental images they can use to help them memorise the facts but also for them to realise how these facts relate back to the structure of the number system.

Other properties of numbers that we introduce to our students are the square numbers. These are also easily represented pictorially, as

The first square number is 1, the second is 4 and so on. We write 1

$$1^2 = 1$$
$$2^2 = 4$$
$$3^2 = 9$$
$$4^2 = 16$$

and so on.

There are also the triangle numbers. 1, 3, 6 and 10 are the first 4 triangle numbers.

You will see that we form the triangle numbers as follows:

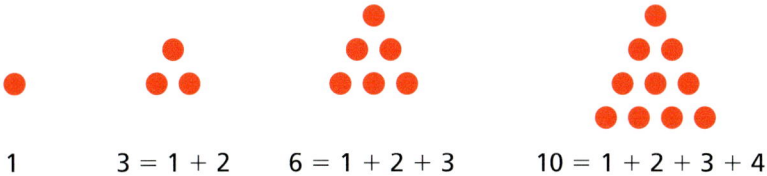

1   3 = 1 + 2   6 = 1 + 2 + 3   10 = 1 + 2 + 3 + 4

and so on.

The next portfolio task asks you to explore the relationship between square numbers and triangle numbers in more depth. In this way you get a better "sense" of these properties.

Portfolio task 5.2

Explore the link between square numbers and triangle numbers. Let us call the square numbers S1, S2, S3, S4, and so on, so that

$$S1 = 1, S2 = 4, S3 = 9, S4 = 16 \ldots$$

and the triangle numbers T1, T2, T3, T4 so that

$$T1 = 1, T2 = 3, T3 = 6, T4 = 10 \ldots$$

Try to represent any square number as the sum of two triangle numbers.

Try to illustrate your answer with a picture.

Rules

The rules that are important to help children in applying their number facts appropriately are the **commutative property** of numbers; the **distributive property** of numbers and the **associative property** of numbers. These might sound complicated, but you will have been using them all your life and, as we have been saying throughout this chapter, they make absolute sense.

The commutative property

This means that for some operations it does not matter which order the numbers are used in the calculation. The numbers can "commute" or change places. You may have already realised that addition and multiplication are commutative. If I am asked to add 15 and 12 I can either work out 15 + 12 or 12 + 15. Similarly, if I want to know 32 multiplied by 13 I can either work out 32 × 13 or 13 × 32. This property is useful as it means we can immediately halve the facts we need to learn. Students are always very happy when they discover this is true for all addition and multiplication equations. The commutative property is also helpful as we can encourage students to rearrange calculations so they can carry out mental calculations. We would use this property if we were to be asked to carry out the following calculation

$$15 + 7 + 14 + 3 + 5$$

as we could rearrange the numbers in the following way

$$15 + 5 + 7 + 3 + 14 = 20 + 10 + 14$$

to make it easier to add the numbers.

It is the commutative law that teachers are using when they suggest to students that they start counting on from the largest number first, when carrying out an addition by counting on a number line. However, the commutative property does not hold for subtraction or division; the order of numbers cannot change. You will see that, for example,

15–8 is not the same as 8–15

and

$$28 \div 7 \text{ is not the same as } 7 \div 28$$

The distributive property

The best way to illustrate the meaning of the distributive property is through an example. Often when we carry out mental calculations for multiplication we rely on this distributive property. For example, the following figure illustrates a calculation using an array.

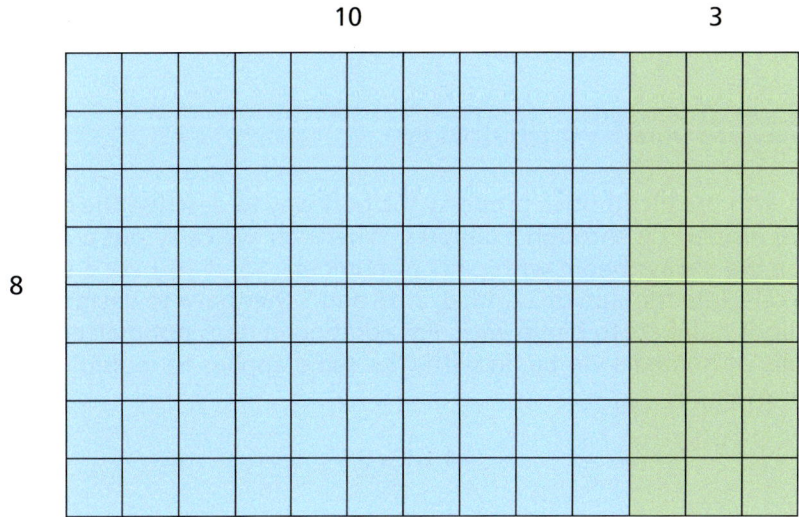

Rather than calculate 13 × 8 I can calculate 8 × 10 and then add 3×8. Then I could write

$$13 \times 8 = (10 \times 8) + (3 \times 8) = 80 + 24 = 104$$

Similarly, to work out 28 × 7 we could write

$$28 \times 7 = (30 \times 7) - (2 \times 7) = 210 - 14 = 196$$

Being able to draw on the distributive property is very helpful in order to develop your mental calculation skills. There are some examples later for you to try out in your portfolio. Always encourage the students you are working with to describe how they are carrying out their mental calculations. Implementing *Number Talks* as part of your regular classroom practice can transform the way your students think about mathematics and develop number sense. Through *Number Talks*, students are exposed to a multitude of pathways and flexible solutions to a single problem. This will help all your students come to terms with the distributive law.

Portfolio task 5.3

Use the distributive property to carry out the following calculations mentally.

Draw a sketch to illustrate the calculation:

18 × 8	34 × 7
29 × 14	48 × 6

Check with a calculator to convince yourself.

The associative property

As with the distributive property the best way to describe the associative property is through examples. Whenever we carry out a calculation we always begin with pairs of numbers. So, if you ask a student to calculate the sum of 22, 14, 6, 3, 18 and 5 they have to decide which pair of numbers to begin with. For addition, it does not matter which pairs of numbers we begin with. The same applies to multiplication. For example,

$$4 \times 7 \times 25$$

or

$$4 \times 25 \times 7 = (4 \times 25) \times 7$$
$$100 \times 7 = 700$$
$$\text{So, } 4 \times 7 \times 25 = 700$$

However, subtraction and division are not associative, for example,

$$18 - 7 - 3 = (18 - 7) - 3 = 11 - 3 = 8$$

but

$$18 - (7-3) = 18 - 4 = 14$$

So, we have to carry out subtraction and division in the order the numbers are given.

These rules are very helpful in reorganising calculations to make them easier but can lead to some confusion as you will see later in the chapter. This is why it is always important for students to develop deep conceptual understandings about why these rules work, rather than just be told that they do. This way, they are more likely to understand the rule and thus not apply it incorrectly. Students should also always check their calculations to make sure they are reasonable and make sense.

The next section of this chapter explores common errors that you might notice students making or misconceptions that they may have developed. The aim of each teaching point is that you are able to use the error or misconception as a focus for teaching to move the students' learning forward.

Teaching points

Teaching point 1: confusion in definitions of properties of numbers

You may have seen questions where students are asked to sort numbers according to their properties either using a Carroll diagram or a Venn diagram. (Examples of Carroll diagrams and Venn diagrams are shown later.) For example, students were asked to sort 2, 7, 8, 9, 17, 20 using this Carroll diagram:

	Even	Not even
Prime		
Not prime		

The main confusion here involved 2 and 9. Children are introduced to odd and even numbers relatively early in primary school and so have a good sense of "oddness" and "evenness," but prime numbers are not introduced until much later. Because all prime numbers are odd, apart from 2, students often make the mistake that all odd numbers are prime, forgetting that 9 has three factors (1, 3 and 9) and so is not prime. Similarly, many children do not see 2 as a prime number even though it has only two factors, 1 and 2.

Games like *Guess my number*, which opened the chapter, are useful for helping students become familiar with the wide range of properties of numbers. Another activity that is useful as a starter is to give students number cards, a different number for each student. Ask them to move around the room, then ask them to stand still and form groups of three. The three in the group then have to find a property that is shared by all the numbers in their group. Another activity that can be repeated as often as you like is to draw a large Venn diagram on the board. One of the class comes up to the front and writes numbers in the appropriate part of the Venn diagram. As they are doing this the rest of the class have to guess what the properties are that are being used for the sort.

Sometimes these misunderstandings can be used directly by asking questions that the students may think are impossible:

Tell me an even number that is also a prime number.

Tell me a multiple of 3 that is even.

Tell me a multiple of 10 that is in the 3 times-table.

And so on.

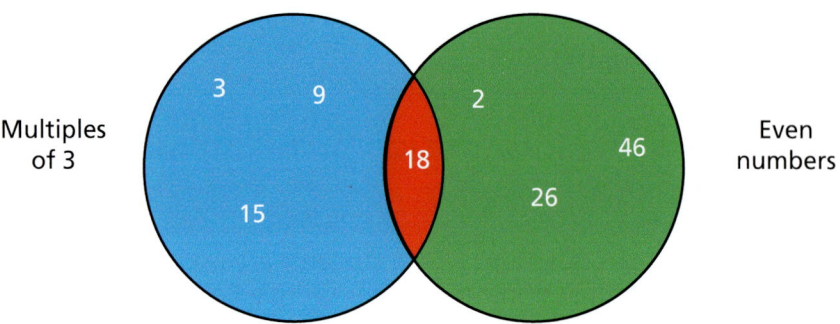

Teaching point 2: errors in remembering multiplication facts

Most of us can remember that every multiple of 5 ends in a 0 or a 5 and that every multiple of 10 ends in a 0. You will probably also remember that all the even numbers are multiples of 2. This sometimes leads to confusion for students who can overgeneralise. They notice that every even number is a multiple of 2 and so invent a new rule, "every odd number is a multiple of 3," which sounds plausible but unfortunately is not true. This is a good example of a misconception.

It is very useful to try to discover the rules that students are following so that you can ask them to find examples that show their self-made rule does not always work. You can provide students with one of their own assumed rules and ask them to prove if it is a rule that is "sometimes" "always" or "never" true. It is also helpful to introduce them to the patterns that do exist for other families of multiples. For example, let us look at multiples of 3:

3, 6, 9, 12, 15, 18, 21, 24, 27, 30

If we add up the digits of any number in the 3 times-table and repeat this until we get a single digit, we always get 3, 6 or 9.

So, 15 gives me 1 + 5 = 6, and 27 is 2 + 7 = 9.

Meaning, 4,341 is divisible by 3 (4 + 3 + 4 + 1 = 12 and 1 + 2 = 3).

Check it on a calculator if you don't believe it. Asking students to explore the patterns in families of multiples is a good way of encouraging them to get a feel or a sense of the properties of numbers.

Portfolio task 5.4

Explore patterns in the 4 and 9 times-tables. Are there rules that you can find so that you know if a large number is divisible by 4 or by 9?

TAKING IT FURTHER: **FROM THE CLASSROOM**

For an example of how a Year 1 teacher explored number patterns using posters see Jill Russell's article "Interactive pictures" in *Mathematics Teaching* 182.

She describes how she used sets of posters to support the children in "telling stories" about the numbers, which in turn developed their understanding of number patterns. There are also examples of the posters in the article. One poster is called "Animal fields" and has pictures of a wide range of animals in fields. Jill Russell asked her class to make up questions based on the poster. One child came up with the following story:

Seven rabbits were just finishing their breakfast when they spotted five sheep in the next field. Four of them ran away frightened so then there were only three of them left.

It is a great idea to ask your learners to make up their own "number stories" based on illustrations or photographs that you give them. This helps them become at ease with numbers and also helps them when they come to interpret number problems that are set for them by other people.

Teaching point 3: errors in estimation when deriving new facts

Sometimes the students get very excited when they realise they can derive a huge number of new facts with the knowledge that they have. A Year 6 group were asked to jot down as many new facts as they could, given their knowledge that $8 \times 7 = 56$. This is the list that one group came up with:

$8 \times 7 = 56$	so	$80 \times 7 = 560$
and $80 \times 70 = 56,000$	and	$800 \times 700 = 560,000$
and $8 \times 0.7 = 0.56$	and	$0.8 \times 0.7 = 0.56$

When they were asked if they were sure they were all correct, they looked a little blank. They had forgotten that they were looking for genuine new facts in their excitement and speed and had just been jotting down numbers that looked right. They checked $80 \times 7 = 560$. They agreed this was correct because, "80 is less than a 100 so the answer should be less than 700." They applied the same logic to 80×70.80 is still less than 100 so the answer should be less than 7,000, as 100×70 is 7,000. Any time we have the students explain their thinking or approach, we are integrating the proficiency of reasoning into their experience. They eventually spotted their error and checked the rest. You might like to do the same.

A useful strategy to get students to look carefully for accuracy is to give them lists like the previous one with errors in them and ask them to find the errors and explain why the incorrect answers are wrong. If they can notice the sorts of errors that can be made, they are less likely to make them themselves.

Teaching point 4: children misapplying the commutative and associative laws

Tony's son, Sam, once brought home the following calculation from school:

$$\begin{array}{r} 584 \\ -\ 376 \\ \hline 212 \end{array}$$

He said, "I know it's wrong, the answer should be 208, but I've done it right." Mentally he had counted on, he knew that 576 was 200 more than 376 so if he added another 8, he would get to 584, but frustratingly for him, using the method he thought he had been taught gave him the wrong answer. This is an example of why some students seem to see mathematics as fraught with confusion: you can do the right thing and still you get it wrong.

What had happened is that Sam had followed an instruction he may have been given by a previous teacher, "you always take the smaller number from the larger number." This rule works fine when you are trying to find differences between numbers (e.g. 6 and 9) but can clearly cause confusion when students are breaking up larger numbers into smaller parts. Similarly, he may have been told, "4 take away 6 doesn't go," so here he was trying to find a way to make it "go" and so he subtracted the 4 from the 6, giving him the answer 2. We need to be very careful with the way we describe methods for carrying out calculations so that we do not unwittingly add to students' confusion.

He may also have remembered the commutative law and thought that it applied to subtraction as well as addition. If you remember, the commutative law says that it does not matter in which order you carry out a calculation. This is fine for addition and multiplication but not for subtraction or division. One positive from this example is the fact that Sam had the knowledge to realise that something had gone wrong with his method for performing the calculation.

Teaching point 5: mental calculation errors in partitioning

The previous chapter introduced you to "arrow cards" or "place value cards." You can use them to support students in developing their mental methods by using partitioning. The following activity shows how place value "arrow" cards can be used to support students in using partitioning to develop their skills in mental calculation.

Partitioning numbers

Use these place value cards.

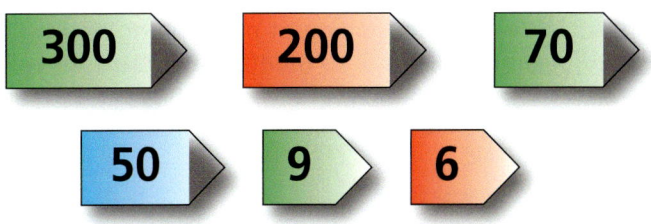

1   how many different numbers can you find by combining these cards?
2   find all the numbers using just two cards each time
3   find all the numbers using just three cards each time.

Here are some examples:

200 + 50 = 250   200 + 50 + 6 = 256

50 + 6 = 56

It is important that students working on this activity are given the cards to manipulate. The overlapping of the cards is a vital part of coming to an understanding of the concept of partitioning. For example, as the student makes 256 by using the cards, they must be able to overlap the card so that they read 256 and not 200, then 50, then 6: this can lead to their developing misconceptions around place value and around partitioning.

Because students realise that partitioning offers them a powerful way of calculating mentally, they sometimes see it as a first port of call rather than one strategy to choose when appropriate. This can lead to students becoming overly reliant on partitioning when other strategies would be more efficient. For example, consider the following equation:

$$49 \times 3$$

When Leo was asked to solve this, he automatically broke the 49 into 40 and 9. So, 3×9 is 27 and 3×40 is 120, thus the answer is 147. However, while 147 is the correct answer, a much easier way of working this out would be to realise that 49 is almost 50 and 3×50 is 150. So, given that 49 is one less than 50, that means I need to subtract 3×1, which also gives me 147.

An overreliance on partitioning can also led to the following misconception. A class had been looking at partitioning as a way of carrying out multiplication calculations mentally the previous week. This week, however, they were looking at division. They were asked to use any strategies they wanted to calculate 280 divided by 14. Sohm had written this in her book:

$$280 \div 14 = 20 + 20 = 40$$

She was asked how she had done this. She said she had partitioned 280 into 200 and 80 and 14 into 10 and 4, and so 200 ÷ 10 was 20 and 80 ÷ 4 was 20 giving her the answer 40. The fact that both answers were 20 seemed to have convinced her even more that it must be right (there's something about round numbers in maths that feels right, isn't there?).

To encourage her to check, she was asked what 40 × 14 was; if her answer was correct this should have been 280. Very quickly, she answered 560. She explained she had used partitioning again. Here she had calculated 40 × 10 and added 40 × 4. This time she had partitioned correctly. She then realised that 280 ÷ 14 could not be 40 and should be 20 (because 20 × 14 = 280). The following examples unpick this in more detail:

$$280 \div 14$$

Partitioning 280 into 200 + 80 does not help as neither of these are easily divisible by 14, and as we saw earlier, splitting this calculation into

$$200 \div 10$$

and

$$80 \div 4$$

gives us a different calculation altogether. For this calculation it is easiest to notice that 28 is double 14 so 280 ÷ 14 = 20. This is an example of how memorising doubles of numbers can help us with mental calculations. However,

$$40 \times 14 = (40 \times 10) + (40 \times 4) = 400 + 160 = 560$$

As mentioned before, a good way to work with the rules of number is always to question your students about the strategies they are using whether they are correct or making errors. This way they will talk to you about the methods they are using; you can pick up the misconceptions and work with them. Very often students are giving you correct answers, they just aren't carrying out the calculation that you asked them to.

In practice

The following lesson plan was used with a group of Year 2 students who had been exploring a range of mental addition and subtraction strategies over the previous few weeks. This lesson was designed to serve two purposes. First, to give the students an engaging way to practise some of strategies they had been learning about. The second aim was to give the teacher time to assess how well the students were able to apply a variety of mental strategies in a game situation.

Following the plan is an evaluation of the lesson that explores how successful the plan was in supporting the children to develop their knowledge, skills and understanding.

Learning intentions

To select the best addition or subtraction strategy to help you solve a range of equations.

To practise using these strategies to improve our fluency.

Key vocabulary

counting on, counting back, tens facts, doubles, near doubles, halves, build to the next ten, adding ten

Resources

Ten sided dice (1–10), Get Out of My House! gameboards, counters

Context for lesson

This lesson was taken mid-way through a unit of work focussing on addition and subtraction. The class had been exploring a range of strategies, including tens facts, doubles and near doubles. This was the first time the class had ever played this particular game.

Warm-up activity: coins in my pocket

The class was told a brief story about the teacher stopping at a café on the way to school that morning to purchase a coffee and receiving 80 cents back in change. That change now sat in the teacher's pocket and the challenge for the class was to guess what coins the 80 cents was made up of. They were told that they had only six guesses to figure it out and that no student could contribute more than one guess.

Randal went first, and he said that the 80 cents was made up of 4 20-cent coins. The teacher recorded this on the board as 20c + 20c + 20c +20c = 80c, and the class checked by counting by 20s to see if this was mathematically correct. The teacher indicated that it was by ticking the equation but also pointed out that this did not match his collection of coins.

The activity continued along this way, with each attempt from a student being recorded on the board and then checked to see if it equalled 80 cents. After three unsuccessful attempts, a further clue was given, with the students being told that there were exactly 8 coins in the teacher's pocket. This led to lots of hands being raised, and the next attempt from Lucy was 8 lots of 10 cents, which recorded as 8 × 10c = 80c. However, as with all of the guesses up until this point, while it did equal 80 cents, it did not match what was in the teacher's pocket.

The fifth attempt came from Cale and it was 20c + 20c + (4 × 10c) = 80c. Another student quickly pointed out that could not be the answer, as it only contained six coins. Finally, Adele said "20 cents plus 20 cents, plus two 10 cents and then add four 5 cent coins," which was correct. The teacher than took the coins out of his pocket to prove that Adele was in fact correct.

Launch

The game that the students were going to spend the lesson playing was called "Get Out of My House!" and it was one that they had not seen before. To introduce it, the class was split into two equal sized groups with about 12 students in each. A large game board was placed up on the magnetic white board

GET OUT OF MY HOUSE!

1	2	3	4	5
6	7	8	9	10
11	12	13	14	15
16	17	18	19	20

YOU WILL NEED: Gameboard showing numbers 1-20 (above)

Pack of cards - 1-10
14 counters (7 of each colour)

HOW TO PLAY: Player 1 selects two cards and adds the two numbers. They place their counter on that number. Player 2 repeats this step. If there is already a counter on the number then all the counters are removed and replaced with their counter. Players may place more than on counter on a number. object of the games is to be the first player to have all their 7 counters on the gameboard.

and each team was assigned a colour and given seven magnetic counters. The teacher got the game started by placing himself on the red team and rolling the two 10-sided dice. He rolled a 7 and a 3. This could be used to make 10 (by adding the two numbers) or 4 (by subtracting 3 from 7). The teacher opted for the 10 and placed one of the 7 red counters in the 10 box.

Next, Ani had a turn for the yellow team. He rolled a 6 and a 5 and used it to make 11 (6 + 5), placing a yellow counter in the 11 box. At this point, the students were told the aim of the game was to get rid of all your team's seven counters. Penny used an 8 and 2 to make 6 (8 − 2) and placed the second red counter in the 6 box.

On the yellow team's next roll, Malakai rolled a 6 and 4 and, at this point, the teacher stopped to explain an especially important rule. He asked what were the two numbers that could be made from a 6 and 4, and the class responded 10 (6 + 4) and 2 (6 − 4). The teacher then explained that if yellow choose to make 10, they could kick the red counter out of the 10 box and shout, "Get Out of My House!" which of course Malakai promptly did.

On the next red turn, Carley rolled two 3s, which she used to make 6. This meant there were now two red counters in the 6 box. At this point, the teacher explained the final important rule. You are welcome to place up to three of your team's counters in any one box. Obviously, when there are two counters in the same box, this is riskier than when there is just one, as the other team can potentially kick two of your counters off the board with just one roll of the dice. However, if you are able to get a third counter in this square, that box is now locked up, meaning that no further counters can be added and the three that are there can no longer be kicked off the board. So, it is a short-term risk for a long-term reward.

The class continued playing the game to its conclusion at which point they were told they would spend the rest of the lesson playing the game with a partner of their choice. Each pair then grabbed a game board, some dice and found a space in the room where they could go and play.

Explore

Students played the game with their partners for about 25 minutes and, during this time, all the pairs remained engaged. This allowed the teacher to roam around the room and observe the students while they were playing. As he moved about, he recorded what he saw in a checklist that he had previously prepared.

For example, he saw Izzy work out 6 plus 2 and 6 minus 2 by using counting on and counting back without any prompting from her partner. In both cases, she was fairly efficient and quietly said the numbers aloud as she was working it out ("Okay . . . six, seven, eight.")

When Luka rolled a 6 and a 5, he immediately placed a counter on 11. The teacher was unsure as to how he worked this out, so he asked Luka to explain. Luka replied, "Well, I know 5 and 5 is 10, so I just added one more to make 11." The teacher then asked a few more near double questions, to check whether Luka could apply this strategy more widely (e.g. "What is 7 and 6? How about 8 and 9?")

The only time the teacher's attention was taken away from assessment was when a few pairs of students approached to say that they had finished playing. They were reminded that once you finished your first game, you were free to play a re-match, which the students were very excited to hear.

Summary

At the end of the lesson, students were asked to write a reflection in their books, responding to the following prompt questions that had been placed on the whiteboard:

- what are some addition/subtraction strategies you used today? Give examples of some equations that you might solve with each strategy
- what have you learned about this game that will help you the next time you play it?
- can you suggest a way to modify the game to make it easier or more challenging?

After a five-minute period of silent writing, students were asked to share their responses. One response to each question was shared with the entire class. When Malakai came up the front, he said he had learned that it is better to use addition if you roll 3 and 9 than subtraction. When asked to explain why, he said, "Well, there's lots of ways to make 6 in this game but it's much harder to make 12." He then proceeded to illustrate his point by recording the different ways to make each number on the board.

Rationale and evaluation

The lesson was extremely successful. The students loved being able to play the game with a partner of their choice and, as a result, there was a very high proportion of on-task behaviour. This meant that the class had close to half an hour in which nearly every student was practising the strategies that they had learnt over the previous weeks.

The structure of the lesson and high levels of student engagement also freed the teacher up, allowing a lot of assessment information to be collected. The fact that this was a planned outcome meant that there was a checklist already prepared, which made it really easy to record observations. This information can now be used to help further planning, as there are some strategies that the bulk of the class seem to have mastered and others where additional time and instruction is likely to be beneficial.

Audit task

Devise a lesson plan that is appropriate for a group of learners you are working with. The focus should be an aspect of knowing and using number facts that is appropriate to the age group you are working with. Use the proforma that is available on the companion website. Make sure you think carefully about the context and evaluate the lesson. Add this lesson plan and evaluation to your subject knowledge portfolio.

Assessing knowing and using number facts

The most effective way of assessing your students' knowledge of number facts is on a day-to-day basis. Playing the *Guess my number* game as in the example that opened the chapter is one way to get a sense of which pupils are at ease with properties of numbers. Similarly, you can choose a number of the week. This can be an individual, group or whole class activity. Simply select a different number each week and during the week the students have to list as many properties or facts as they can about the number.

Loop cards are a great way of assessing your students' current knowledge of number facts as well as supporting them in learning new number facts, through noticing patterns and hearing other children's responses. Loop cards are a set of cards on which the answer to the question on one card leads to the next card; so for example one card might read, "I am 15% of 200," and another card will have the answer "30" written on top of the card and a new question on the bottom of the card. You can also see how they are using current facts to work out new ones by stopping the game and asking students how they arrived at an answer.

There are further examples of loop cards and a blank set on the website that accompanies this book.

Cross-curricular teaching of knowing and using number facts

The idea of these extended cross-curricular projects is to draw on the key ideas within the chapter to develop a cross-curricular project that you can explore with your learners over a series of lessons. This allows you and your class to develop your subject knowledge together.

Boxes for stock cubes

This investigation asks students to explore rectangular prisms with constant volume. This involves them in looking at prime numbers, factors and multiples. It is most effective for students to work in small groups. Give each group enough multilink cubes to make a range of rectangular prisms made from 36 cubes. You will need to encourage the students to keep a record of the rectangular prisms they make – for example, 3 × 4 × 3, and so on. They will probably notice that rectangular prisms with the same dimensions look "different" in different orientations.

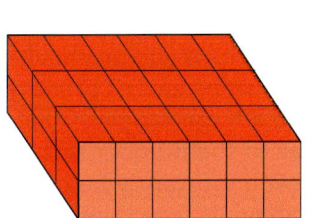 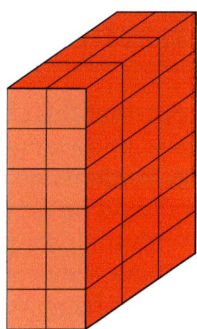

Ask groups to explore the range of rectangular prisms that can be made from different numbers of cubes; you could suggest they work systematically on this starting with one cube, then two cubes, then three, then four, and so on. They can then decide what "sort" of numbers give the largest number of different rectangular prisms. They will discover that numbers that have a large number of factors can be made into a large number of rectangular prisms, and prime numbers can only make a single rectangular prism.

This leads them into exploring boxes of stock cubes. Bring a range of stock cube boxes into the class. Ask the groups to decide which rectangular prisms they would use if they were packaging 36 stock cubes. They should create the packaging for this and prepare a "bid" for the rest of the class to argue for their design. The whole class can judge the best designs as a concluding activity for the investigation.

Summary

The chapter opens with a discussion of number sense. The development of this number sense is an underpinning theme for the chapter, and we described it as being at ease with numbers or trusting the internal logic of the number system. The progression section outlines which number facts are appropriate for each stage of learning, but the chapter also focuses on techniques you can use to support students in using the facts that they already know to learn new facts. Using patterns to remember and predict number facts is illustrated, and the rules that underpin the number system are carefully outlined. You will also have seen how this supports students in coming to an understanding of algebra. The teaching points illustrate how you can use students' misconceptions around number patterns and the rules of number to bring them to a better understanding of how the number system works. If they trust the number system they will also trust their memory.

Reflections on this chapter

We hope that this chapter has helped you see how you gradually build students' knowledge of number facts over seven years of primary education. Students do not have to memorise a huge range of facts. The focus should be on students making sense of the number facts associated with their stage of learning and most importantly helping them to see how they can use the facts that they already know to derive new sets of facts. You should now understand which facts you should help your students learn and which you should be working towards and understand the difference between fluency, number sense and memorisation.

In terms of "using number facts" this chapter has aimed to explain why certain rules work. We hope you understand the commutative, associative and distributive properties of numbers and how the misapplication of these rules can sometimes lead to errors. Finally and most importantly, we hope you are beginning to see that the number system makes sense and it is feeling less mysterious and more logical. In the foundation stage you teach your students how to add 1 to a number below 10, so they understand that every time you add 1 to 8 you get 9. Armed with this knowledge you can carry out an infinite number of calculations, eventually you know that because $1 + 8 = 9$, then $100 + 800 = 900$, $0.1 + 0.8 = 0.9$ and $28.01 + 3.08 = 31.09$. What could be more logical than that?

Carry out this self-audit to examine how your learning has progressed as a result of working on this chapter. Include your results in your portfolio.

Self-audit

This activity asks you to use your knowledge of place value and multiplication facts to multiply 10 by 10 to derive related facts involving decimals.

I know that 8 × 7 = 56 so I can derive the facts 56 ÷ 7 = 8 by rearranging the number sentence; 28 ÷ 7 = 4 by halving 56; 2.8 ÷4 = 0.7 by dividing by 10.

Start with 12 × 10 and write down 10 other number facts you can work out from this starting point. Explain how you derived each new fact.

1

2

3

4

5

6

7

8

9

10

Going further

Fluency without fear

JO BOALER

Fluency without fear is available on the youcubed website at www.youcubed.org/fluency-without-fear/

Making number talks matter: developing mathematical practices and deepening understanding, grades 4–10

CATHY HUMPHREYS AND RUTH PARKER

This book introduces you to *Number Talks*, a 15-minute classroom routine that supports students to use reasoning in explaining their mental maths strategies. The book outlines the different strategies students will use to perform any operation and provides helpful examples that can be used in the classroom.

Humphreys, C. and Parker, R. (2015) *Making Number Talks Matter: Developing Mathematical Practices and Deepening Understanding, Grades 4–10*. Portland, ME: Stenhouse Publishers.

The number mysteries: A mathematical odyssey through everyday life

MARCUS DE SAUTOY

This is a book to engage and challenge you in learning more about number facts for yourself. It is a very accessible text and hopefully will fascinate you. You will be able to draw on much of what you learn from this book in your own classroom, and it will certainly make you much more confident – and excited about the properties of numbers.

De Sautoy, M. (2011) *The Number Mysteries: A Mathematical Odyssey Through Everyday Life*. London: Fourth Estate.

A proposed framework for examining basic number sense

ALISTAIR MCINTOSH, BARBARA REYS AND ROBERT REYS

This article introduces the term "number sense." The authors explore the link between informal methods and formal calculation procedures. Ian Thompson analyses the concept further in the paper "Narrowing the gap between mental computational strategies and standard written algorithms" presented to the International Convention on Mathematics Education in Denmark in 2008.

McIntosh, A., Reys, B.J. and Reys, R.E. (1992) 'A proposed framework for examining basic number sense', *For the Learning of Mathematics*, 12(3), 2–8.

CHAPTER 6

NUMBER
Calculation (addition, subtraction, multiplication and division)

Introduction

Try asking your students to think of someone they know who is good at mathematics. Then ask them what this person can do that makes them good at mathematics. One student said recently, "They can't just do additions, they can do subtractions too"; another said, "They can do really hard things like long divisions." An answer that is often given is, "they can work things out really quickly, in their head and they don't even need to use a calculator." Although the main thrust of this book is to illustrate that mathematics is much more than simply calculating, this suggests that for many of us our image of mathematics is linked predominantly to calculating.

This chapter will explore how you can develop your capacity to, "Do it really quickly in your head." Developing your own mental strategies in order to help you teach your students to calculate mentally can be challenging as you may not have been encouraged to do this when you were learning mathematics. Some students we have talked to remember being taught a particular method that seemed to work most of the time but remember times when the method did not seem to work (see previous chapter). Because they only had one method to fall back on, they were unable to carry out the calculation, particularly in high-pressure situations like tests or examinations.

This chapter will introduce you to the idea of "efficient and effective methods." This is a move away from teaching one method for all calculations. You are efficient and effective at calculating when you have a range of methods to draw on and can select the most appropriate method for the particular calculation you are working on. You will also come to accept that the method that is most appropriate for

DOI: 10.4324/9781003315155-6

one person may be different for another person. Through discussing our own methods, we can expand the range of methods we have to draw on.

A key section in this chapter illustrates how important using concrete materials can be when introducing written methods to learners. Manipulatives like bundling sticks and base-10 blocks can be used to model calculations. Using these materials models for students why the algorithms they are introduced to, to calculate using pencil and paper, such as column addition and subtraction, actually work.

You will also see how addition, subtraction, multiplication and division all connect. If you can add, you can subtract; if you can multiply, you can divide. Hopefully by the end of the chapter you will feel confident that you are a good calculator. Perhaps this will go a long way towards convincing you that you can be good at mathematics too. This will give you the confidence to use your calculating skills in solving problems, an important functional skill.

Starting point

Work out the answers to the following calculations mentally. Avoid writing anything down at this point:

$$11 + 6 =$$

$$28 - 9 =$$

$$20 \times 5 =$$

$$57 \times 3 =$$

Now ask yourself, "how did I carry out the calculation?" You may think, "I just did it." But let us think more deeply.

Did you just know that 11 and 6 made 17, or did you add 6 and 1 to make 7 then add that to 10? Did you partition 9 into 8 + 1 so you could take 8 away from 28 to get 20 and then take away another 1 to give 19?

Did you multiply 10 by 5 and then double the answer to get 100? Did you add 20 + 20 + 20 + 20 + 20, or is this a fact you know?

And finally, did you work out 3 × 50 and 3 × 7 and then add 150 + 21 to give you 171, or did you use the algorithm, or rule, you were taught in school? In your head you might have said something like, "3 × 7 = 21; put the 1 down and carry 2; 3 × 5 is 15, add 2 is 17. So, the answer is 171." This models the written algorithm.

When you carried out these calculations you were making decisions about what was the most effective strategy for you. It is unlikely that you drew on one formal written method for any of these calculations.

In Chapter 5 you were introduced to the idea of number sense. It was this number sense that you drew on to decide how to come to an answer. The only time this number sense might have deserted you is if you drew on the formal algorithm to carry out the multiplication. You may have lost sight of the value of the numbers, saying something like, "add 2 to give me 17 and then another 1, gives me 171." This describes the patterns of the numbers rather than their values, which can lead to confusion.

TAKING IT FURTHER: **FROM THE RESEARCH**

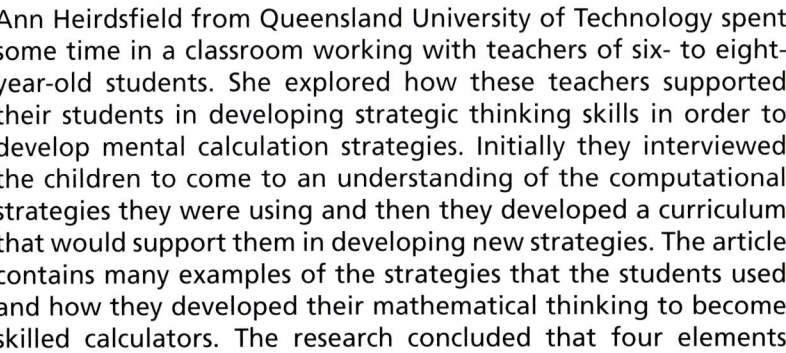

Ann Heirdsfield from Queensland University of Technology spent some time in a classroom working with teachers of six- to eight-year-old students. She explored how these teachers supported their students in developing strategic thinking skills in order to develop mental calculation strategies. Initially they interviewed the children to come to an understanding of the computational strategies they were using and then they developed a curriculum that would support them in developing new strategies. The article contains many examples of the strategies that the students used and how they developed their mathematical thinking to become skilled calculators. The research concluded that four elements were present that allowed students to develop efficient and effective mental strategies. These were:

1   It is important to determine students' existing knowledge. This is best done through discussion, through asking, "What strategy did you use to calculate?"

2   Connect areas of mathematics so that students see how strategies they use for one calculation can be transferred to other calculations.

3   Teach concepts associated with the strategies so that students can see the connections. Representations such as 100 squares and number lines are helpful here.

4   Maintain a classroom environment in which students feel safe to explore, share, critique and justify the strategies they are using.

As you can see, there is much more to calculating than just knowing the written methods for adding, subtracting, multiplying and dividing. The message is clear: mental strategies first, then a range of written methods to allow children to develop "efficient, reliable and effective" methods when it is not possible to use mental strategies.

Chapter 13 looks at the use of calculators in detail and describes how you can introduce these to your students so that they are supported in developing their mental strategies and know when it is appropriate to use them. In this chapter you will read about the importance of developing a range of calculating methods so that your students can make sensible choices over their methods when they are faced with a calculation to carry out.

TAKING IT FURTHER: **FROM THE RESEARCH**

In Chapter 4 you were briefly introduced to *Children and Number: Difficulties in Learning Mathematics* (1986, Blackwell) by Martin Hughes. It is worth revisiting this research at this point. There is an important section exploring calculation strategies and the way in which children invent their own forms of recording when carrying out calculations. Hughes suggests that, left to their own devices, children go through a series of stages in their responses.

An initial stage is the *pictographic* response. At this stage children literally draw the calculation. So, if they are adding three multilink blocks to four multilink cubes, they will draw the cubes in two groups and then count them up to add them together. Another stage is described by Martin Hughes as the *iconic* stage. Here children represent the blocks by icons. For the previous calculation, a child might draw three sticks in a circle and four sticks in a circle and then add them. Again, there is a direct correspondence between the drawing and the physical act of counting. Both of these stages precede the use of formal symbols; here a child would write the symbol "3" and the symbol "4" to represent the calculation.

I have seen children using the number symbols as icons – for example, when working out

$$3 + 4 =$$

they will count each point of the 3 by tapping their pencil on the points and then tap each point of the 4 to carry out the calculation. They count the taps and reach 7. You may have seen children doing this – you may even remember doing it yourself!

Take the time to spend some time with early learners in your setting as they start to record their calculating process. Encourage them to use their books to jot down their working out. It is likely that you will notice them developing in the way that Martin Hughes suggests.

Progression in calculation

Foundations for calculation

Counting is at the heart of calculating, and initially you will be working with your learners on counting objects and using language such as "more" and "less" to introduce them to ideas of addition and subtraction. You should compare different groups of objects to find out how many more or less there are in each group and start to explore addition and subtraction using concrete materials in practical situations.

Beginning calculation

The next step is to introduce your students to the idea of counting on. That is, if you are adding 5 and 3 you point to 5 on a number line and count on saying, 6, 7, 8. You can teach that addition can be carried out in any order and that you can carry out subtraction by finding the difference between two numbers through "counting up." This also shows them how addition and subtraction are inverse or opposite operations and the two operations should be taught together. You may introduce the number symbols to record practical activities but should ensure students are secure in their understanding before you do this.

Students will come to recognise multiplication as repeated addition (one of those connections we talked about earlier) and will represent this using arrays. For example,

4 × 3 is the same as

You will also be using concrete materials to represent division as grouping into equal sets, and your students will use these representations to solve simple division problems.

Becoming confident in calculation

At this stage you can begin to introduce children to the four symbols, +, −, × and +, as well as =. This will help children find unknown numbers in number sentences such as 20 + __ = 12, which is the early stages of algebra (see Chapter 7). Mentally they will be able to add and subtract:

- a two-digit number and ones
- a two-digit number and tens
- two two-digit numbers
- three one-digit numbers (adding only).

They will use concrete objects and pictorial representations to support them in this. Children should also understand that addition and subtraction are **inverse operations** and should represent multiplication as an array –

At this stage you can support your students to add or subtract mentally one- and two-digit numbers and introduce them to practical and informal methods to multiply and divide two-digit numbers (these informal methods are described later in the chapter), teaching them that multiplication and division are inverse operations. Your students will also start to find unit fractions. Unit fractions have 1 as the numerator. So, they will be able to find 1/2 of 5 metres or 1/6 of 18 litres, for example.

You can develop your students' use of written methods to record and explain multiplication and division of two-digit numbers by single-digit numbers and teach them to use written methods to add and subtract two- and three-digit whole numbers, including working in the context of money. You should expect your students to multiply and divide by 100 and 1,000, and they will have developed their understanding of fractions to find fractions of quantities and shapes. You will explore the use of a calculator to help them carry out one- and two-step calculations involving all four operations, including understanding the meaning of negative numbers on the display screen of the calculator

The next stage is to develop calculation skills so they can use efficient written methods to add and subtract whole numbers and decimals up to two decimal places. You will have helped them extend their mental strategies for whole number calculations including multiplying one-digit numbers by two-digit numbers and subtracting near multiples of 1,000. You will ask them to draw on their understanding of place value to multiply by 10, 100 or 1,000 and be able to draw on a range of written methods to multiply and divide three-digit numbers by single-digit numbers. They will be able to find fractions and percentages by division including using a calculator.

By the end of primary school, students will be calculating mentally when appropriate and will be able to use a range of written methods for those calculations which cannot be carried out mentally. You will expect them to use a calculator to solve multi-step problems and find fractions and percentages of whole number quantities.

Extending learning in calculation

With these students you can teach them to apply the commutative, associative and distributive laws that were described in Chapter 5, as well as using their understanding of inverses to calculate more efficiently. Look at these examples:

$$12 \times 13 = 156$$
$$13 \times 12 = 156$$
$$156 \div 12 = 13$$
$$156 \div 13 = 12$$

For example, multiplication and division are inverse operations; this means they are the opposites of each other. By knowing the answer to one of the previous problems you can work out all the others.

The students will have developed their mental strategies to include fractions and decimals and will be able to calculate percentage increases and decreases.

Big ideas

To introduce the big ideas in calculating try this activity:

Portfolio task 6.1

Work out 48 + 36 and record your solution.

When using this activity with teachers the most common response is

$$\begin{array}{r} \overset{1}{4}8 \\ + \ 36 \\ \hline 84 \end{array}$$

When asked "how" they calculated the answer a common response is, "I added 40 and 30 to give me 70 and then added on 14." What is interesting here is that this calculation is not represented by the method they record. The written method records the question and the answer but does not represent the method used. When working with students, they are likely to use a number line to record the calculation (if they have been shown this method). For example

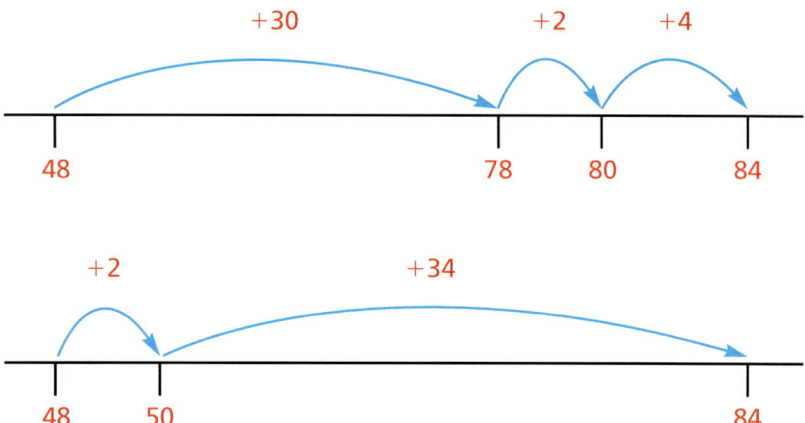

The big ideas in calculating are how students develop mental strategies, which support them in understanding and developing formal written methods at an appropriate stage. Another big idea is how to use concrete materials and manipulatives to support the development of written methods.

Students' development of mental strategies

The reason that we encourage students to develop mental approaches to calculation is because it is much more efficient to work things out mentally if we can, rather than immediately reverting to a pencil and paper or a calculator. If we are fluent in mental strategies, we carry this confidence over into other areas of mathematics. The reason to develop rapid recall of number facts is to help us become fluent in mental calculation, which will be beneficial in many other areas of our lives, not so that we can pass our weekly times-tables tests.

This is just the same for number facts and mental calculations. The most appropriate strategy to support children in developing their mental strategies is to use a wide range of images and models that children can draw on to enable them to develop mental images of calculations. Using number lines, 100 charts and empty number lines all support the development of mental strategies.

Activities such as *Guess my number* or *Loop cards*, discussed in the previous chapter, give you plenty of opportunity to ask individuals how they are calculating mentally and also support your students in developing effective mental strategies. You should join in these activities too and make your own thinking overt, sharing it with your learners. In this way you will make sure that you develop your own mental capacity too.

Students develop mental calculation skills as they progress through school. At each stage they should be using concrete materials and images and models of the number system to support them. The following table illustrates the progression in developing mental calculation skills and the mental images that you can use to support students.

Mental calculation skill	Strategies and models
Adding and subtracting pairs of single-digit numbers Adding and subtracting single digits from 20	Use concrete objects – grouping together and removing from groups Use number lines and number tracks
Adding single-digits numbers to and from multiples of 10 Adding and subtracting single digits from two-digit numbers	Number lines Hundreds charts Empty number lines Partitioning
Adding and subtracting groups of numbers Adding and subtracting two-digit numbers from two-digit numbers	Reorder to make calculation simpler Number lines Empty number lines Hundreds charts Partition numbers
Adding and subtracting three-digit numbers and 1s and 10s and 100s Multiplying two-digit numbers by one-digit numbers	Empty number lines Use known facts to derive new facts Partition numbers
Multiplying three single-digit numbers Multiplying numbers by near multiples of 10 and 100	Empty number lines Use known facts to derive new facts Partition numbers
Using place value, known and derived facts to multiply and divide mentally	Empty number lines Use known facts to derive new facts Partition numbers

You will see that as the mental calculation becomes more complex, we have to draw on an important skill. That is, we need to be able to use facts that we know to derive new facts. This is best described through an example. By the later stages our students will have learnt that we can carry out mental multiplication by partitioning.

$$28 \times 7 = (20 \times 7) + (8 \times 7) = 140 + 56 = 196$$

I know that

$$28 \times 7 = 196$$

Using this fact, I can list the following facts

$$280 \times 7 = 1960$$
$$2.8 \times 7 = 19.6$$
$$0.28 \times 7 = 1.96$$

and so on.

Try this for yourself.

Portfolio task 6.2

Use mental strategies to carry out the following calculations

$$36 \times 9$$
$$52 \times 6$$
$$84 \times 3$$

Derive at least five new facts from each calculation.

Students' development of written methods

There has been a long-held belief that young children cannot carry out calculation activities until they reach what Piaget called the "concrete operational thinking" stage at around age seven. Martin Hughes, who carried out the research explored earlier in the chapter, challenged this view suggesting that very young children can begin to explore the underlying processes of calculating. He developed a "box task." In this task he worked with children aged between three and five. With some children he asked them to count a number of bricks into a box, which he would then close. He would then either add additional bricks to the box or remove bricks and ask the children how many were in the box now. The young children could understand and succeed on the task so long as the numbers were very small.

When working with one child he counted five bricks into the box, then removed three and asked the child how many were left. The young boy told him "two." Martin Hughes then recorded the following conversation (see page 27 in his 1986 book *Children and Number*):

MH: I want to take three bricks out of the box now.

Richard: You can't can you?

MH: Why not?

Richard: You just have to put one in, don't you?

MH: Put one in?

Richard: Yeah, and then you can take three out.

This suggests that Richard was capable of carrying out two successive mental calculations before he was five. Asking students to represent activities such as this early in their experience in school will help them see written methods as supportive of mental calculations rather than simply a record of the question and the answer.

TAKING IT FURTHER: **FROM THE CLASSROOM**

In the article "Good concrete activity is good mental activity in the journal, Australian Primary Mathematics Classroom," Andrea McDonough describes how important concrete materials are for young children to develop mental calculations strategies. As with the research discussed earlier in the chapter, the research showed how important the teacher is in supporting young learners in using concrete materials efficiently. Whilst recognising that access to concrete materials is beneficial, Andrea McDonough suggests that activity should be planned that explicitly uses concrete materials and that it is important that teachers intervene (or teach) appropriately to develop students' skills in using the materials.

The next section of the chapter draws heavily on a series of articles written by Ian Thompson for the Association of Teachers of Mathematics (ATM). These appear in *Mathematics Teaching*, issues 202, 204, 206 and 208 and explore in detail an approach to calculation that has been drawn on by many effective schools. These are available from the ATM and can be found online at www.atm.org.uk/.

Addition

An initial approach to written calculation is the empty number line that we introduced you to at the beginning of the "Big ideas" section. The empty number line is a useful "record" of a mental strategy as you can use it to support any addition or subtraction calculation. We could view the conceptual structures for addition in the following way:

1 Combining two or more quantities

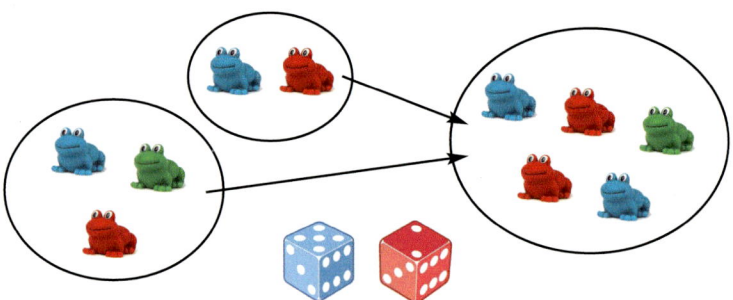

2 Augmentation of one quantity

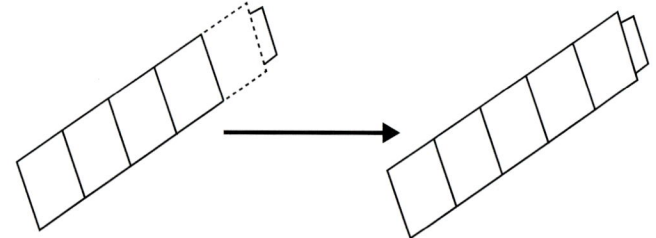

3 Comparative addition

3	2
5	

The use of this bar model has recently been encouraged as it is seen to be a part of the Singapore method for teaching methods. Indeed, textbooks that have been written to encourage a Singapore approach focus heavily on the bar method. It is an important image for children to have and a useful tool for calculating. However, as you can see from all the images and representations offered here, it is simply one of a range of possible models that we can use.

4 Using a number line.

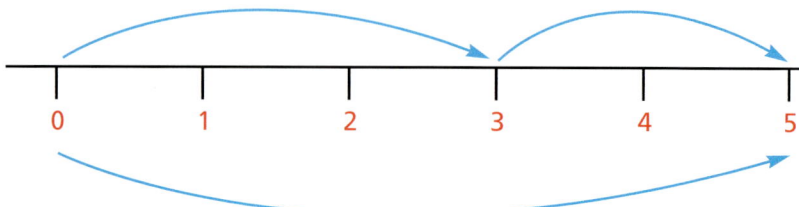

The next idea to introduce is the use of **partitioning** for calculating with larger numbers. For example,

$$39 + 52 = 39 + 50 + 2 = 89 + 2 = 91$$

or

$$39 + 52 = 30 + 50 + 9 + 2 = 80 + 11 = 91$$

Can you see that only the first example of partitioning builds on the use of an empty number line? If you draw a number line you can sketch the calculation on that line. The second example changes the order of the numbers so that an empty number line would not model the calculation as you need to conserve the order of the calculation to sketch it in a number line. The second form of partitioning can then be represented in a more formal way as

$$
\begin{array}{rl}
39 \quad = & 30 + 9 \\
+ 52 & \underline{50 + 2} \\
& 80 + 11 = 91
\end{array}
$$

This leads students into the expanded column method. This method would record the calculation above as either

$$
\begin{array}{l}
39 \quad \text{(adding 10s first, that is, 30 + 50)} \\
\underline{+52} \\
80 \\
\underline{11} \\
91
\end{array}
$$

or

$$
\begin{array}{l}
39 \quad \text{(adding 1s first, that is, 9 + 2)} \\
\underline{+52} \\
11 \\
\underline{80} \\
91
\end{array}
$$

It is helpful to work your examples both ways so that your students can see there is no difference to adding 10s first or 1s first. Before teaching the column method involving trading, students need experience regrouping numbers, for example, demonstrating that 142 can be decomposed into 1 hundred, 4 tens and 2 ones but can also be represented as 0 hundreds, 14 tens and 2 ones. When introducing column addition or subtraction, ensure you do so with the assistance of place value materials such as MABs. In order to make sure that students make the links back to place value it is suggested that they are encouraged to say "trading 10" or "trading 100" rather than "trading 1" as may have become habit.

Subtraction

We can view subtraction in two ways: subtraction as an action (i.e. taking something away) or as the difference when we are comparing two things. Sometimes students will model a subtraction calculation such as 15 – 6 by placing 15 objects in a box and taking out 6 and then counting how many remain. This will give them 9. They may represent this by a jotting like this:

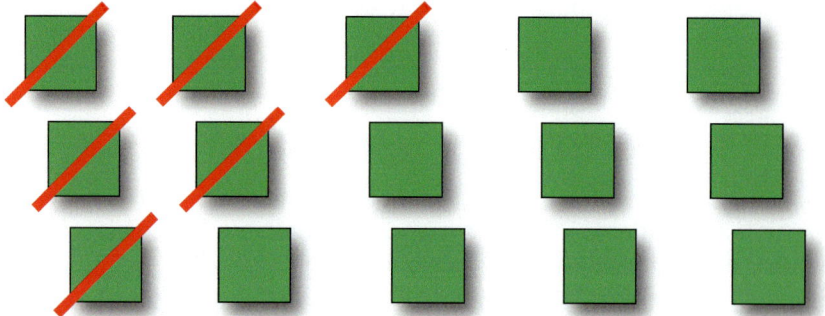

Subtraction as difference can be calculated using a number line. Here the student will start at 15 and count back 6 ending up on 9.

A way of thinking about the conceptual structures for subtraction is as follows

1 Partitioning

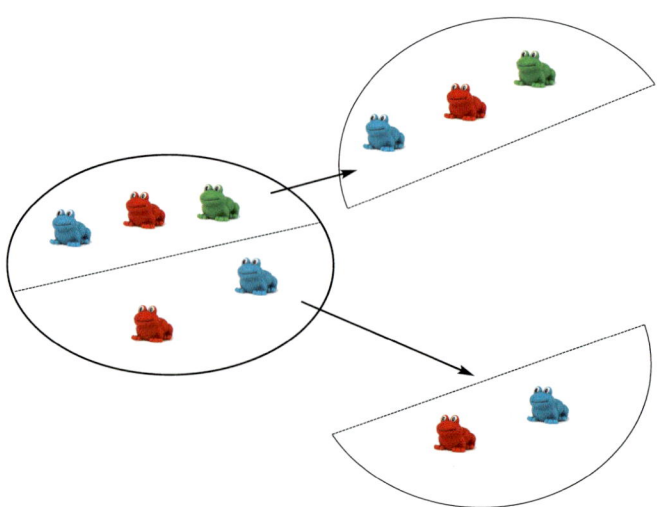

but adapt so that there are three frogs of one colour in one half of the circle and two frogs of a different colour in the other half.

2 Reduction

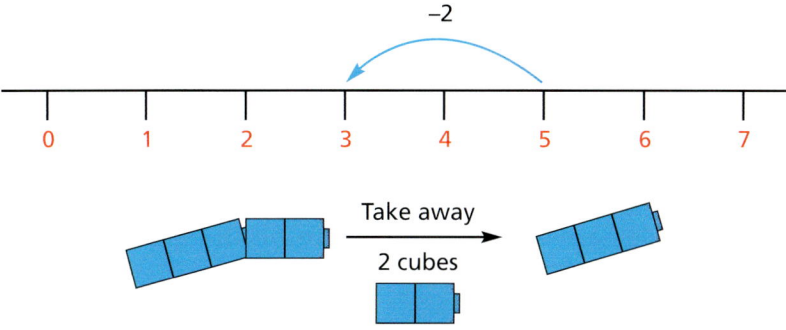

3 Comparative difference

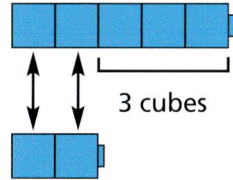

4 Additive difference/complementary addition.

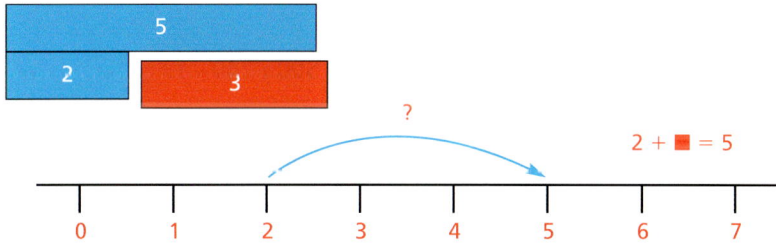

Again, the bar method is illustrated as a step in between working with concrete materials such as cubes and moving towards using a number line to model subtraction.

The stages for developing written methods to support subtraction also employ a progression through the empty line to partitioning before moving on to column methods. The first stage in the strategy involves counting back on an empty number line. So, for example, to calculate 74 − 27 we may write

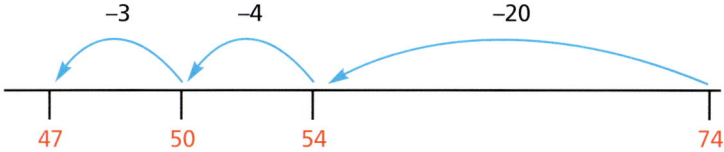

145

or combine steps by initially subtracting 7 and then 20. It is important that students see the calculation can be carried out in any order.

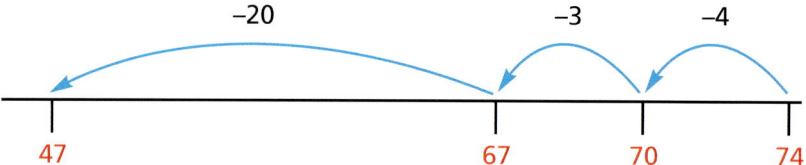

An alternative is to begin with the 27 and ask the students how many they would have to count up to get to 74. This can be recorded on an empty number line as follows:

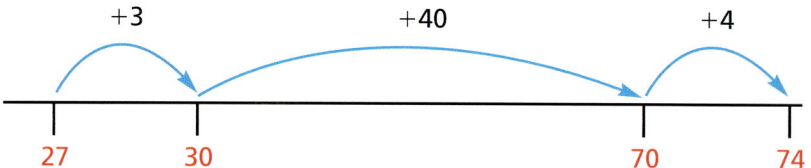

Ian Thompson suggests that if students can use the empty number line in this way there is no benefit in using partitioning to support subtraction. In fact, students are effectively partitioning if they are using the empty number line efficiently.

The final stage in moving towards a formal algorithm is the expanded layout. An example of this is 653 − 289:

$$
\begin{array}{r}
600+50+3 \\
-200+80+9 \\
\hline
\end{array}
\qquad
\begin{array}{r}
500+140+13 \\
-200+\ 80+\ 9 \\
\hline
300+\ 60+\ 4
\end{array}
$$

In moving from one calculation to the other we have partitioned

$$
\begin{array}{rcl}
600 & = & 500+100 \\
50 & = & 40+10 \\
3 & = & 3
\end{array}
$$

So, 653 can be written as 500 + 140 + 13.

Multiplication

Ian Thompson suggests it is important to let students develop their own informal methods to record multiplication as with addition and subtraction. Students need to first explore the connection between repeated addition and multiplication, securing the new concept by connecting it to their prior knowledge and experience. As students become more comfortable repeatedly adding groups of objects or images, it's important that they begin organising these items into arrays. An array is a way of arranging items into rows and columns, allowing students to see a visual representation of multiplication, also forming a steppingstone to using the area model.

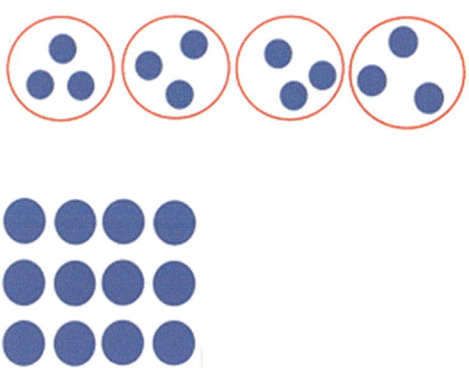

The area model can be used to introduce students to the ideas of the traditional algorithm via a more visual representation. It is worth spending some time becoming secure with these methods yourself so that you are able to introduce them confidently to your students. Begin by introducing it with simple one-digit by two-digit problems using concrete materials such as MABs or the squares on grid paper to create a partitioned array. This will tap into their prior knowledge about arrays and will consequently make better sense of the area model.

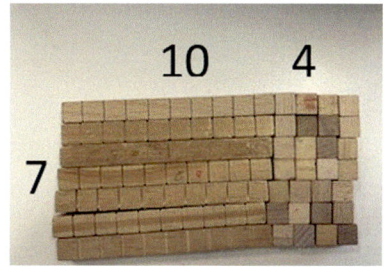

This can be represented using grid paper in the following way

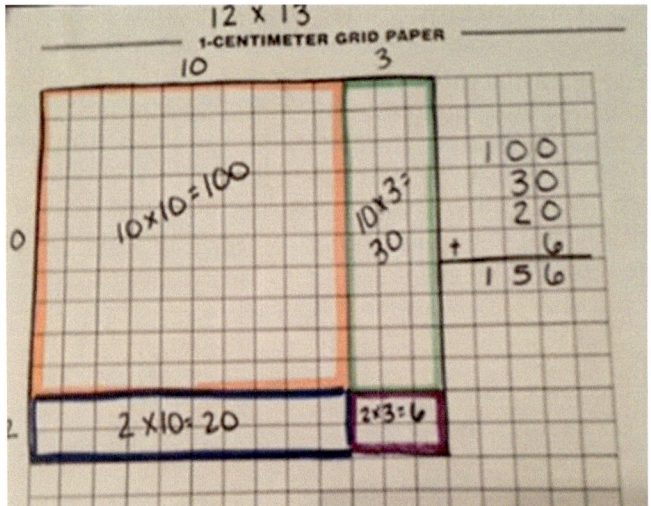

The next step is the area model, which is more of a short-handed approach where you are continuing to retain the relative proportionality of the numbers in the problem but giving less regard to the actual rows and columns.

We would write 38 × 7 as:

38 × 7

	30	8
7	210	56

210 + 56 = (266)

The area model would then be extended to multiply larger numbers like a two-digit number by another two-digit number. For example, 29 × 37 would look like:

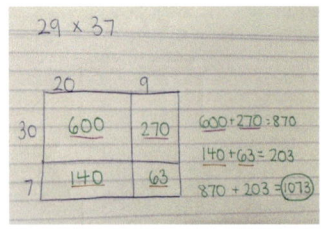

29 × 37

	20	9
30	600	270
7	140	63

600 + 270 = 870

140 + 63 = 203

870 + 203 = (1073)

A formal method that makes sense of the traditional algorithm is called the partial product strategy. This approach bridges the more visual representation of the area model with the numerical representation of the traditional algorithm. It looks like this:

$$
\begin{array}{r r l}
324 & \color{green}{(300 + 20 + 4)} \\
\times \quad 16 & \color{green}{(10 + 6)} \\
\hline
24 & \color{red}{6} & \color{green}{\times 4} \\
120 & \color{red}{6} & \color{green}{\times 20} \\
1800 & \color{red}{6} & \color{green}{\times 300} \\
+ \quad 40 & 10 & \times 4 \\
200 & 10 & \times 20 \\
3000 & 10 & \times 300 \\
\hline
5184 &
\end{array}
$$

It is only when students have worked through these progressive steps to build their number sense should you introduce them to the traditional algorithm or "long multiplication." Ultimately, this method is a process for a quick calculation with a lot of room misconception and error if it is taught with haste.

$$
\begin{array}{r}
{\scriptstyle 1\ 2} \\
32\!4 \\
\times \quad 16 \\
\hline
1944 \\
3240 \\
\hline
5184
\end{array}
$$

TAKING IT FURTHER: **FROM THE CLASSROOM**

In the article "Simplifying and sorting computation strategies," published in the journal *Prime Number* in April 2020, Dr Judy Harnett suggests that strategies for computation can be characterised as either a **breakup** method, or a **change** method. In a breakup method one or both numbers in a calculation are broken up into parts often based on place value. For a change method the numbers may be changed whilst maintaining equivalence. For example, the following would be "break-up" methods:

$$6 \times 125$$

	100	+ 20	+ 5
6	600	120	30

$$600 + 120 + 30 = 750$$

Example 1: Place value parts

$$4 \times 27$$

	25	+ 2
4	100	8

$$100 + 8 = 108$$

Example 2: Compatible parts

$$4 \times 137$$

	100	+ 30	+ 5	+ 2
4	400	120	20	8

$$400 + 120 + 20 + 8 = 548$$

Example 3: A combination
These are examples of change strategies

$$9 \times 48$$

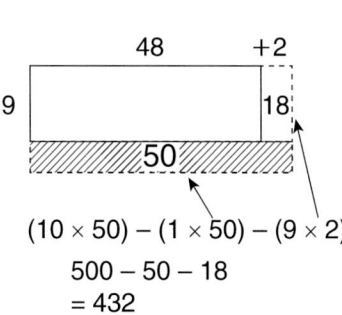

$$(10 \times 50) - (1 \times 50) - (9 \times 2)$$
$$500 - 50 - 18$$
$$= 432$$

Example 4: Changing two numbers, counting extra once.

$$9 \times 48$$

$$(10 \times 50) - (1 \times 50) - (10 \times 2) + (1 \times 2)$$
$$500 - 50 - 20 + 2 = 432$$

Example 5: Changing two numbers, counting extra twice.

As you work with students ask them to try to illustrate their strategies pictorially and see if they are examples of "break-up" strategies or "change" strategies or maybe a combination of both.

Division

When we talk to people about why they do not feel confident about teaching mathematics they often describe "long division" as a hated memory. If we want to ensure that this is not the case for the next generation of students, it is important that we explore division in a different way. Our aim should not be to teach students a process ("long division") that they memorise, with very little understanding about what they are actually doing. Instead, we should focus on building concept understanding through exploring a variety of strategies for dividing.

$$
\begin{array}{r}
2\,7 \\
3\,\overline{)\,8^21}
\end{array}
$$

Children have an intuitive understanding about equal sharing as it is an experience they have all had, whether it be sharing food with siblings or toys and cards with friends. It is important that students are encouraged to draw on what they already know about division through hands-on learning opportunities with manipulatives and later on through drawings. They will most likely begin by distributing the dividend by ones, one at a time. In solving the equation 21 ÷ 3, students might collect 21 counters, for example, and dispense them into 3 different groups, one at a time, like this:

With multiple exposures to this sort of task and teacher-modelling, students will gain confidence and begin realising that they do not have to distribute by ones but instead use their knowledge of common skip-counting patterns. Given the same problem as earlier, you would expect the students' thinking would progress to something that looks like this with more of a numerical representation:

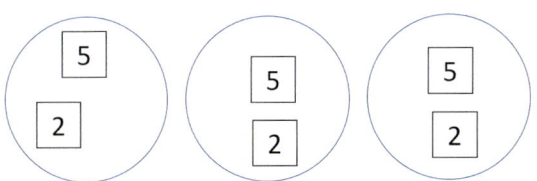

It is important that students develop an understanding of division as the inverse of multiplication. For example, you can use a number line to show a range of strategies for division (see the following examples).

Empty number line strategies for division:

1 Using repeated subtraction to solve 24 ÷ 6

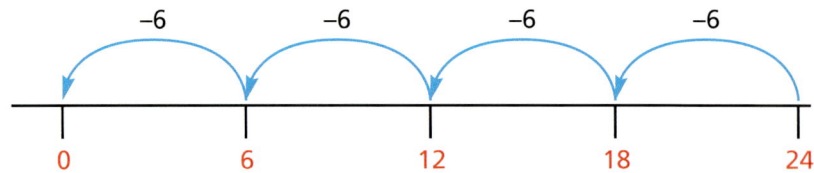

So 24 ÷ 6 = 4, because 24-6-6-6-6 = 0

2 By "chunking" to solve problems with larger numbers where repeated subtraction becomes less efficient. This strategy relies on an understanding of multiples. For example: 96 ÷ 6.

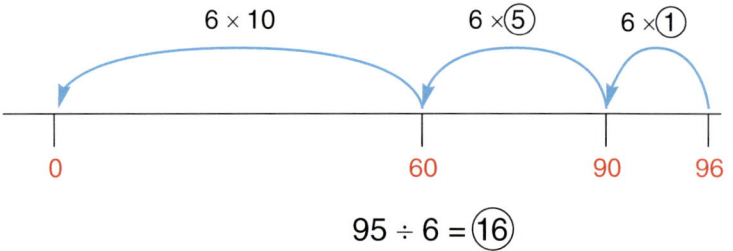

$$95 \div 6 = \circled{16}$$

The first chunk was 6 × 10, which brought us to 60. If we had done that same chunk again, we would have over-shot our target, so the next chunk was halved, (6 × 5) bringing us 30 closer to 96. With only 6 remaining, we knew it was only one more group of 6 to reach our target.

It is also important not to rush into any formal recording of division. We still hear students in schools or pre-service teachers describing this process in the following way, "81 divided by 3. So, 8 divided by 3 goes 2 carry 2; 3's into 21 goes 7. So, the answer is 27."

The problem with this form of calculation is that the actual place value of the numbers is forgotten: 80 becomes 8 and 20 becomes 2. This can lead to students not noticing if answers are incorrect, as they do not return to think if 27 is a sensible answer to 81 divided by 3. Estimating first can help this process. We would expect the answer to be less than 30 because 90 divided by 3 is 30. We would also expect it to be more than 25 as 75 divided by 3 is 25. The answer should therefore be between 25 and 30, so 27 seems about right.

Initially students can use partitioning to carry out division calculations. For example, if they are calculating 75 divided by 3, the dividend of 75 is split into 60, the highest multiple of 3 that is also a multiple 10 and less than 75, to give 60 + 15. Each number is then divided by 3. So,

?	?
60	15

3 | 60 | 15 |

$$75 \div 3 = (60 + 15) \div 3$$
$$= (60 \div 3) + (15 \div 3)$$
$$= 20 \div 5$$
$$= 25$$

This short division method can be recorded as,

$$20 + 5$$
$$3\overline{)60 + 15}$$

which can be shortened to 25.

As students become more confident, they can move onto division with larger numbers, including those that result in a remainder. For example, to calculate find 226 ÷ 6, we might start by multiplying 6 by 10, 20, 30 . . . to find that 6 × 30 = 180 and 6 × 40 = 240. This means that the answer to 226 ÷ 6 is somewhere between 30 and 40.

So, we can start the division by first subtracting 180 from 226 leaving 46. We know that 7 × 6 = 42. We have,

$$6 \times 30 = 180$$
$$6 \times 7 = 42$$

Represented on an empty number line, it would look like this:

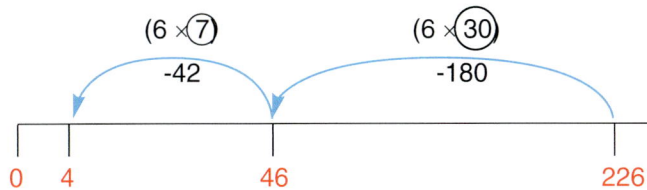

Or through the area model:

30	7
180	42

6 | 180 | 42 | R4

Adding gives us 222 (there will be a remainder of 4). This can be written as

$$226$$

$$\frac{-180}{46} \quad (6 \times 30 = 180)$$

$$\frac{42}{4} \quad (6 \times 7 = 42)$$

Answer: 37 remainder 4.

When dealing with remainders, it is important to conceptualise them with your students. A task we like called "Sharing 25" from *Mathematics Assessment for Learning: Rich Tasks and Work Samples* by Downton, Knight, Clarke and Lewis, prompts students to share 25 balloons between 4 people, then $25 between 4 people and finally 25 biscuits between 4 people. As mentioned earlier, equal sharing is intuitive to children and most of the time, they see a need to share fairly. The items in this task contextualise the remainder, allowing them to be dealt with in different ways. As students become more familiar with fractions and decimals, they can also begin to deal with the remainder, trying to divide it down as far as possible. We would therefore expect students to solve the problem 226 ÷ 6 as earlier but with a quotient of 37 4/6 or reduced down further to 37 2/3 for a more precise answer.

Using manipulatives and practical resources to support calculation

All teachers are happy to use practical materials to aid calculation at the earliest stages. We would think it was silly to take away practical materials from a student who was just beginning to combine sets to understand addition or taking objects out of a set to begin to understand subtraction. Unfortunately, we often reduce their use as students move through school even though they can act as incredibly useful models to help us understand calculation strategies and in particular written calculation strategies. However, it is clear that students will benefit from using manipulatives both to model mathematical processes (particularly calculation) and to give them mental images with which they can visualise mathematical operations.

TAKING IT FURTHER: **FROM THE RESEARCH**

In 2017, Joseph Turner and Nancy Worrell explored the history of using manipulatives in learning and teaching mathematics and offered advice on how to use them effectively. In their paper *The importance of using manipulatives in teaching math today*, they reminded us that manipulatives, on their own, do not teach mathematics or lead students to make mathematical discoveries but that teachers play a vital role in supporting students to use manipulatives. They also suggested that the most important factor in the effective use of manipulatives was that teachers were trained in their use. We would encourage you to explore the use of manipulatives further and share how you use these materials as a school. The paper closes with a very helpful list of manipulatives that are most commonly used in mathematics classrooms today.

The three resources that we think are most helpful in supporting students in learning to calculate are Dienes material, or base-10 material and Cuisenaire rods. We would hope that every classroom would have all these resources freely available.

Diene's material (Base-10 blocks)

These blocks, made up of *ones, tens rods*, *hundreds flats* and *thousands cubes* are widely credited to Zoltan Pal Dienes, a Hungarian mathematician. They were widely used in schools across Australia.

These are invaluable in modelling addition and subtraction as you can physically exchange 10 ones for a ten rod. Similarly, it is clear that 10 ten rods make 100 and then 10 one hundred flats make 1,000. If I were to model 33–15, I could use blocks in the following way.

A little research on the internet will lead you to lots of videos that illustrate how to use base-10 blocks. I would recommend that you explore some of these with your students, whatever age they are. We have also found that asking students to show their parents how to use these resources at parents' evenings or workshops is a great way to show parents why the algorithms they learnt as children actually work.

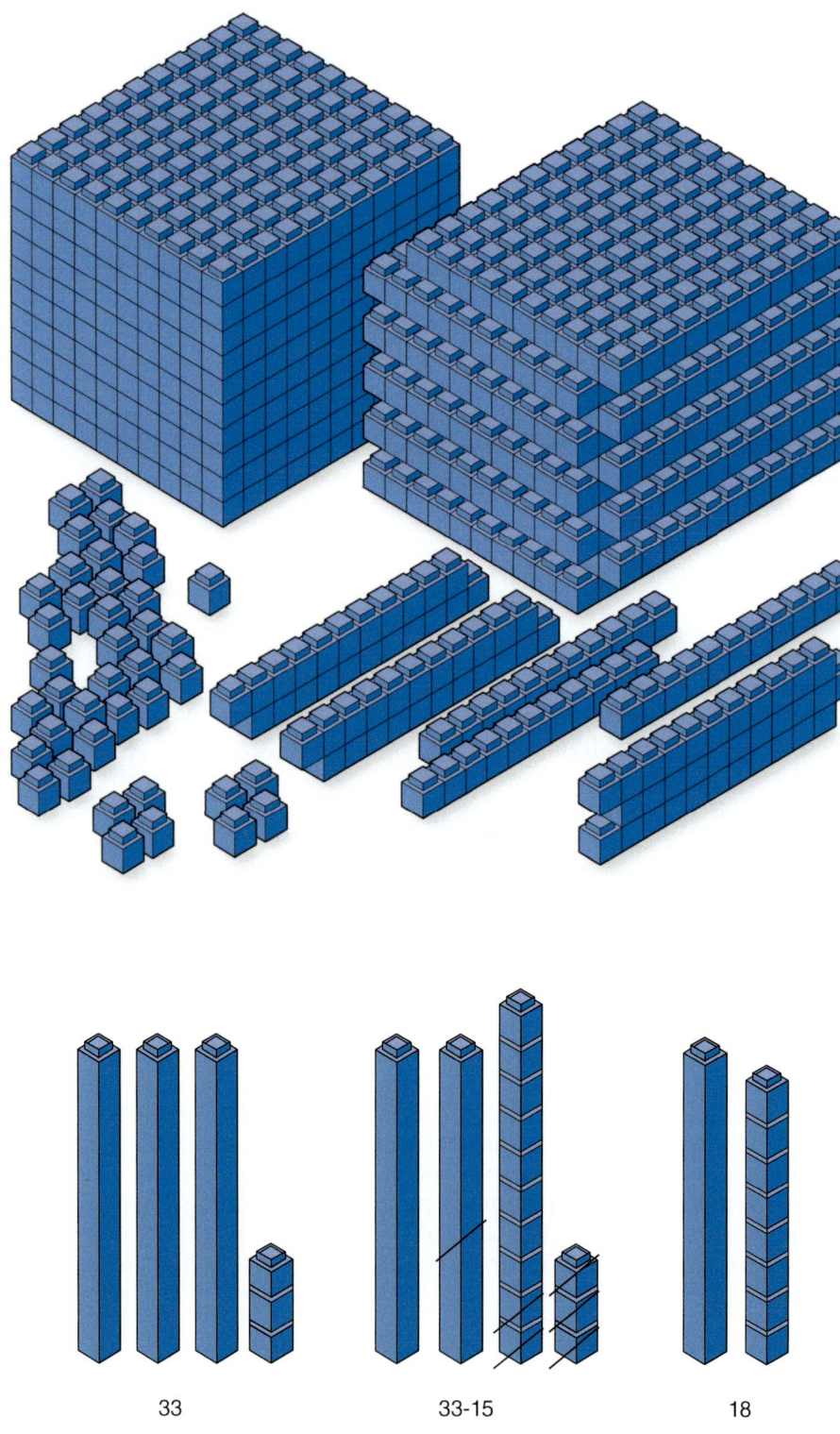

33   33-15   18

Cuisenaire rods

When this image is used with teachers, they often start to make notes of the fractions that they see. Almost always using the orange rod as 1. This can be extended by asking what they see if the orange rod represents 2 or 20 or 100 and so on. This is the power of the rods. They offer a representation of the numbers in relation to each other. In this way they were the forerunner of the bar model. Interestingly, when the image is used with young students they often begin to use algebra. Even though they do not realise this is what they are doing. They might write:

$$Y + Y = O$$

$$R + R + R + R + R = O$$

Which can be interpreted as, "Two yellows make an orange" and "Five reds make an orange." These are very often the first equations these students have written.

The rods were popularised by Caleb Gattegno in the 1950s and were invented by Georges Cuisenaire, a Belgian primary teacher. Gattegno founded the *Association for the use of Teaching Aids in Mathematics* in 1952. So, using manipulatives in mathematics is not new!

As with base-10 blocks, Cuisenaire rods can be used to carry out and model a wide range of calculations. For example:

$$5 + 8 = 13$$
$$13 - 8 = \ 5$$

$$8 \times 3 = 24$$
$$4 \times 6 = 24$$
$$6 \times 4 = 24$$

Cuisenaire rods can also be used to carry out calculations with fractions. This visual representation helps students understand the algorithms for carrying out calculations with fractions.

TAKING IT FURTHER: **FROM THE CLASSROOM**

In "Mathematical journeys: Our journey in colour with Cuisenaire," published in *Mathematics Teaching 257*, Jenny Cane describes how the rods are used by every child in her school to develop their understandings of calculation. Every child in her school is given their own set of rods on entry to the school. They keep this set with them through the whole of their primary education. Jenny explains how they developed Gattegno's own textbooks to become the guiding resource for the mathematics curriculum in the school. She finishes the article with a quotation we still find inspiring:

The journey so far has been exciting and inspirational. There is much more to do and further progress to be made, but we have a strong sense of direction. We truly believe that we are inspiring our children and not simply informing them.

It could be argued that we can see the resources described above as a move through the three stages that Martin Hughes suggested. Base-10 material offers a pictorial representation. We can see the value of each of the blocks; place value counters offer an iconic view and Cuisenaire rods allow us to operate in the abstract as we change the values of the rods.

Place value counters can also be used to model and carry out calculations. For example,

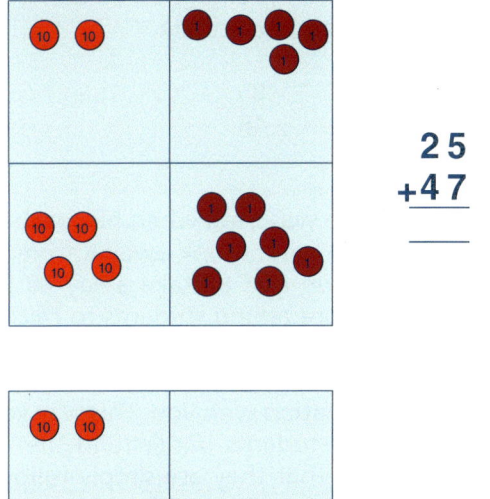

$$\begin{array}{r} 25 \\ +47 \\ \hline \end{array}$$

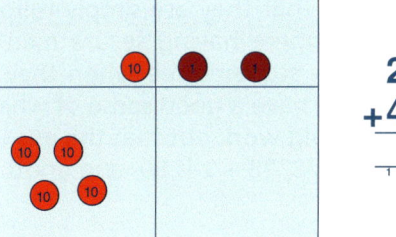

$$\begin{array}{r} 25 \\ +47 \\ \hline 2 \\ \tiny{1} \end{array}$$

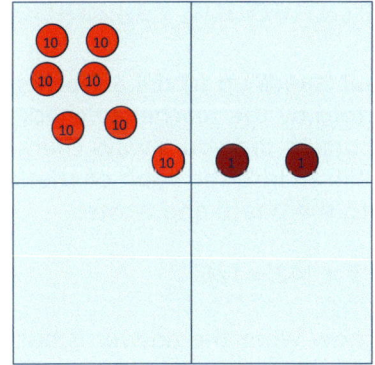

$$\begin{array}{r} 25 \\ +47 \\ \hline 72 \\ \tiny{1} \end{array}$$

Teaching points

Teaching point 1: misapplying algorithms

We would argue that even when giving you an incorrect answer, sometimes students have not 'made a mistake." They might have come up with an answer that does not seem quite right, but when you ask them what they have done, they have answered a question correctly. It just is not the question you have asked them. For example, Sam, aged nine, had been asked to calculate

$$
\begin{array}{r}
85 \\
- 48 \\
\hline
\end{array}
$$

and gave the answer 43. He was irate when his teacher suggested this was wrong; he replied, "It isn't. I took away 8 from 5 by finding the difference, so I put 3 down and 80 take away 40 is 40 so that gives me 43." Whenever you are asking students to calculate using a formal algorithm it is useful to have them talk it through, sometimes at the front so that the rest of the group can see the approach, and sometimes in a one-to-one situation with you. This way you will begin to understand whether your students understand the mathematics behind the algorithms or whether they are simply following or misapplying a rule. This is also where having secure mental strategies is helpful as this can help the students see when they have misapplied an algorithm as they will have a good sense of what the answer should be. In this case Sam could work out that the answer should be 37. He counted on, 48 + 30 is 78, 78 + 2 is 80, and another 5 is 85. So the answer is 30 + 5 + 2 = 37.

Teaching point 2: not checking calculations

We all remember putting our hands up to tell a teacher we had finished our work, only to be told by the teacher to check our answers carefully. Be honest with yourself, did you always check carefully? It is important to try to get children into the habit of checking as a first step. When Helen came up to the board and wrote:

$$3 \times 142 = 126$$

Her teacher said, "I don't know what the answer is but I know that it isn't 126." Helen looked at the teacher and then said, "Oh no, it should be more than 300, shouldn't it?" She was then able to find her mistake and carry out the calculation. Sometimes, when students are following algorithms, they lose track of the numbers they are working with and so forget to estimate and use this estimate as a check. By encouraging students to estimate as a first step you will help them get into this habit.

In the activity below the students are being encouraged to look for errors by drawing on their knowledge of odd and even numbers. It is important that the students are not asked to carry out the calculation.

1   Will the results of these calculations be odd or even?

   a) 4861 + 3758   **b)** 7052 – 507   **c)** 34 × 57

   d) 711 – 296   **e)** 72 × 28   **f)** 461 + 836

2   Which of these are incorrect? Use rules for odd and even numbers
   to check.

   a) 30 × 91 = 2730   **b)** 9072 – 978 = 8272

   c) 826 + 7095 = 7922   **d)** 68 × 47 = 3196

3   Which of these are incorrect? Use rules for odd and even numbers
   to check.

   a) 76 + 128 = 204   **b)** 241–129 = 111

   c) 97 – 38 = 58   **d)** 50 + 503 = 553

4   Will the results of these calculations be odd or even?

   a) 419 + 282   **b)** 174 – 86

   c) 347 + 63   **d)** 70 – 47

Teaching point 3: not understanding written symbolism

The research by Martin Hughes described in some detail earlier in the chapter suggests that often students' errors in calculations are not because they do not understand the mathematical concepts, rather they get confused with how the written symbolism relates to the calculations they are carrying out. It is a problem of translation rather than a problem of mathematics. The best way to support students in making this translation is to work with them as they say number sentences and to write calculations down for them. In this way they will begin to see the + sign as simply shorthand for saying "add."

Another activity to help students begin to understand the use of the multiplication sign "×" is to introduce them to arrays. An array is an "orderly arrangement of objects." Arrays can help us understand multiplication by lining up objects in rows and columns. So, for example, 4 × 3 can be shown as

Looking at a box of eggs, we can see both 6 and 3 × 2.

Similarly, for a box of a dozen eggs we can see 12 and 2 × 6.

If we buy a book of stamps arranged in the way they are in the image below we can see both 24 and 3 × 8.

This sense of multiplication as an array helps both with multiplication and with rapid recall of number facts. It is worth drawing on the many examples of arrays that your students see around them every day. The milk bottles in the crate; cupcakes in a baking tray; lining up in pairs. Making physical arrays is a great activity. Get a group of 12 or 24 of your class and see how many different arrays they can form. Ask the students to make sketches of these arrays.

Teaching point 4: carrying out the wrong operation when solving number problems

Tony recently picked up a greeting card and contained the following text:

> "Jack," said the teacher, *"you have 30 sheep in your field, two escape through a hole in the wall, how many are left?"*
>
> *"None,"* Jack replied instantly.
>
> *"No Jack, it's 28!"* the teacher replied sounding cross, *"I thought you knew your maths!"*
>
> *"Yes Miss, I know my maths alright, it's you that doesn't know sheep."*

This is a great example of a student solving a mathematics problem in their "actual" real world than the pseudo-reality of a maths problem. Students who become successful solvers of word problems realise that these are very rarely "real" problems but are mathematics problems set in a context. A skill for your student to develop is to decode such problems so that they apply the correct operations to solve the problem and then check the answer.

It is helpful to just focus on a series of questions without solving them to teach your students to select the operation which will solve the problem. Spend some time simply underlining key pieces of information in questions in order to help them see through all the extraneous information that is contained in word problems. Focusing on selecting the correct operation without having to carry out the calculation allows your students to see the importance of this step.

Another useful classroom activity is to provide your students with a sketch or an image from which they can devise their own word problems. For example, an image of a cup containing six pencils with two pens lying by the side of the cup will allow students in the Early Years to devise several different word problems. The answers might be 6 + 2 = 8 or 8–2 = 6 or even 2 + 2 = 4 (there are two pencils outside my cup, I take another two out how many are outside the cup now?). If you visit www.gapminder.org/teaching/materials/ you will find many images illustrating global issues that can be used for your learners to devise number problems with the added bonus of developing their global awareness.

Portfolio task 6.3

The area model: Calculate the following using the area model. Before you carry out the calculation estimate the answer. Which two numbers do you think the answer should be between?

358 × 34

×	30	4	
300			
50			
8			

Then use the partial product method shown earlier in the chapter. Can you see how the partial product method is simply a contracted version of the area model?

In practice

The following lesson plan was used with a group of Year 5 students. Following the plan is an evaluation of the lesson that explores how successful the plan was in supporting the children to develop their knowledge, skills and understanding.

Learning intentions

- Develop and use a variety of mental and written methods to solve a range of multiplication equations

Key vocabulary

product, estimate, rounding, distributive property, partitioning

Context for lesson

This way taught midway through a series of lessons on multiplicative thinking. In previous lessons, a number of different methods for multiplying multi-digit equations have been introduced and explored with the students.

Warm-up activity: number talk

Number Talks, developed by Ruth Parker and Kathy Richardson, are 10–15-minute-long mental maths routines designed to engage students in meaningful mathematics discussions. The students are posed with an equation to solve in their heads then encouraged to share aloud the strategies they used to find the solution.

There are routines as part of the structure specifically designed to ensure a safe and comfortable atmosphere where students feel confident to explain their thinking. Students do not raise their hands when they have solved the problem. Instead the students hold their fist against their chest, using a thumbs up signal to privately indicate to their teacher that they have solved the problem. This also means that they are willing to explain their strategy to the class once enough thinking time has been given.

The equation chosen for this lesson was 28 × 4, and straight away they were onto it. After only 30 seconds, about a quarter of the class was giving the thumbs up while the rest were staring at the board with intent.

After giving them another minute or so, Alex shared his strategy, "I knew 28 was made up of 2 tens and 8 ones, so I partitioned it to make it easier. I multiplied 20 by 4, which was 80, then multiplied 8 by 4 which was 32. I then added the 80 and the 32 together." On the board, this is what I recorded: (20 × 4) + (8 × 4) = 80 + 32.

Charlotte said that she also used partitioning but decomposed the numbers differently. Her strategy was recorded as: (25 × 4) + (3 × 4) = 100 + 12. I had anticipated that someone would share this approach and therefore give me an opportunity to make the comparison between the two strategies that perfectly illustrated the distributive property (which we then added to our vocabulary list).

Eduardo chose a different method, explaining that he used the compensation strategy: (30 x 4) – (2 x 4) = 120–8. We then discussed the term "compensation" and made sense of the meaning behind the new term, adding it to our vocabulary list.

Launch

Place the following equations on the whiteboard or up on a screen:

1   27 × 83
2   256 × 8
3   87 × 23
4   46 × 52
5   78 × 32

Students are asked to get themselves into groups of two or three and are then instructed to estimate which equation on the board will have the largest product. Tell them that they only have three minutes, and they won't have access to pencils, paper or any devices. Thus, they will need to use their estimating skills, rather than trying to work out the answers exactly.

At the end of the three minutes, ask a member of each group to tell you which equation they thought had the largest product. Don't ask for any explanation of why at this point, just quickly go around the room and record one of the group member's name next to their chosen equation.

Now, explain to the class that their job for today's lesson is to go back and work out the answer for each equation, to check whether their estimate was correct. Inform them that they can use any of the methods we have looked at over the previous few lessons.

Explore

Students spend time working individually or in pairs working out the solution to all five equations.

Summary

The class was brought back together and the correct answer for each equation was shared. For some equations, simply recording the solution on the board is fine. However, try to select two or three students who can come up the front and demonstrate to the class how they found the answer, talking through the method that they used.

Rationale and evaluation

The students are always engaged in our Number Talks as this is a regular part of the classroom routine.

The way you record the students' thinking is crucial. The board looked like this at the end:

$$(20 \times 4) + (8 \times 4) = 80 + 32$$

$$(25 \times 4) + (3 \times 4) = 100 + 12$$

$$(30 \times 4) - (2 \times 4) = 120 - 8$$

Visually, it makes it very easy to then compare and make connections between the different approaches. It is important to plan the Number Talk beforehand, anticipating the possible strategies that the students will share. Think about how you are going to illustrate their thinking in the clearest way possible prior to revealing it to the students.

When I planned to use this equation, I knew it wasn't going to be too difficult for this group. My intention was not necessarily to challenge them. Depending on the students' strategies, I was either going to focus my attention on the distributive law or doubling and halving. As no one offered me the latter, we explored the former. I noted to myself that the following week I would give a similar problem, and if no one came forward with a doubling and halving strategy I would offer it myself.

In the main activity, students were motivated to find the solutions to the five equations, as it was linked to determining whether their initial estimate at the start of the lesson was correct. Most students used a method based around partitioning the numbers into place value parts. And it was these methods that I thought would be most beneficial to highlight during the summary phase of the lesson.

Ryan was one student who was asked to share his work. He was challenged to justify how he knew his method for calculating 256 × 8 had produced the correct answer. Ryan said that he had simply broken 256 into three parts, 200, 50 and 6, and then he had multiplied each of these parts by 8. So, 8 lots of 6 was 48, 8 lots of 50 was 400 and 8 lots of 200 was 1,600. He then demonstrated on the whiteboard how he recorded this.

The main lesson was a great way to provide students with a purpose for practising their estimation and calculation skills.

Assessing calculation

We would argue that "problem solving" is at the heart of learning calculation skills. This supports the argument presented to you in Chapter 3 and suggests that the ability to solve problems should also be central to assessing your students' calculating skills. This can be carried out in a wide range of contexts. It is helpful if these contexts are genuine "real-life" contexts rather than the pseudo reality we referred to earlier in the chapter. Examples that you can use could include:

- calculating the best value purchase. Bring in a range of cereals or drinks in different sized containers. The students have to calculate which size offers the best value
- calculating the costs to redecorate the classroom. Include the costs of paint for the walls and carpet for the floor
- calculating costs of fruit for younger students. Make sure you let the students devise their own ways of recording their calculations at this stage. It is likely that these learners will begin with repeated addition and will sketch the fruit they want to buy and "attach" a price to it.

Ask very young learners to work out how many raisins there are in a packet. Observe how they group the raisins. Some may choose to count in 2s or 3s or even 5s.

Cross-curricular teaching of calculation

The idea of the extended project is to draw on the key ideas within the chapter to develop a cross-curricular project which you can explore with your learners over a series of lessons. This allows you and your class to develop your subject knowledge together. The examples given above in the assessment section all offer cross-curricular routes into calculating. We have found that developing the assessment task around a hands-on project that brings in real world data a very fruitful activity that involves all learners in developing their calculation skills.

This series of lessons involves analysing and weighing the amount of waste created by your class each day over a whole week.

Have the students develop the different categories that the waste will be sorted into, for example, compostable food scraps, paper/cardboard, soft plastics, etc. At the end of each day, the students weigh the waste from each category.

The resulting data from these weigh-ins can be used in a multitude of ways. Before beginning and throughout the week, students should be involved in estimating the amount of waste. They can calculate the total waste each day by combining the weight of each category. From one day to the next, the students can compare and find the difference between the weight of each category. Once they have a feel for how much waste is created each day, they can then begin calculating the amount of waste across all of the classrooms at the school. They can use these calculations to work out the approximate amount of waste diverted to recycling as opposed to landfill.

Using a cross-curricular approach creates learning opportunities that will meet both the needs and interests of an array of students across a range of year levels.

Summary

The key focus for this chapter is the link between the mental strategies we employ as a first choice and how these can be formalised into written algorithms when the calculations become too complex for us to hold in our head. You have read how the four operations link together and how you can use your knowledge of one of the operations to support calculations using another operation. Our aim for students leaving school is that they have a good understanding of the four operations and that they can carry out calculations mentally, when possible. Students should also be able to employ a range of written methods or make use of devices such as calculators when necessary, making sensible choices depending on the calculation they are carrying out.

We have emphasised the importance of children approaching calculation practically and through activity – this allows you to support them in developing the vocabulary of calculation. They should also be encouraged to develop their own informal records for calculations. The key to understanding the more formal algorithms is to build carefully on previous understanding so that your students understand the reasons that the algorithms "work" and don't just apply them unthinkingly. Most of the misconceptions come from losing sight of the value of the numbers in the calculation so that children don't notice they have an answer that doesn't make sense.

Work your way carefully though the progression of written methods – the area method may be an idea that you are not used to. If the written methods make sense to you, you will be able to share this with your learners.

Reflections on this chapter

We wonder if many of you will have found this chapter the most challenging in the book, not because the ideas are more complex but because the methods of calculation may well be different from those that you were taught. Our hope is that you can see that the rationale for teaching these methods of calculation is that students will be able see why the methods are effective and will begin to move away from simply trying to remember how you carry out a calculation to understanding why a particular method works. If you understand how the calculation is carried out, you can make effective decisions about which method to use and are also more likely to notice when you have made a mistake.

We also hope that you have become able to notice the choices that you are making when you carry out calculations, whether mentally or on paper. In a sense you have "relearnt" the algorithms for the four operations, noticing, maybe for the first time, why you carry out calculations in a particular way. This process of revisiting your own learning should support you in explaining methods of calculation – both mental and written – to your students.

Going further

Teaching mental computation strategies in early mathematics

ANN HEIRDSFIELD

This paper, referred to earlier in the chapter, describes research in two primary classrooms to develop a curriculum that supported young students in developing mental computation strategies. It contains lots of examples of strategies and how best to model these in the primary classroom.

Heirdsfield, A. (2011) 'Teaching mental computation strategies in early mathematics', *YC Young Children*, 66(2), 96–102.

The importance of using manipulatives in teaching math today

JOSEPH FURNER AND NANCY WORRELL

This paper, referred to earlier in the chapter, examines the manipulatives most commonly in use in the classroom. They emphasise the importance of teacher taking the lead and modelling the use of the manipulatives for students and that all teachers should undergo training in the use of manipulatives.

Furner, J.M. and Worrell, N.L. (2017) 'The importance of using manipulatives in teaching math today', *Transformations*, 3(1). Available at https://nsuworks. nova.edu/transformations/vol3/iss1/2

Issues in teaching numeracy

IAN THOMPSON

This title contains a very useful section on calculation that outlines progression in written and mental calculation in much greater detail than there was space for in this chapter. There is also a very helpful chapter on using the empty number line that you will find invaluable in supporting your learners in developing mental strategies.

Thompson, I. (2010) *Issues in Teaching Numeracy in Primary Schools*. Buckingham: Open University Press.

Teaching and learning early number

IAN THOMPSON

This is the forerunner of the previous text. It is important for all teachers, not just those who teach learners in the Early Years, as Thompson outlines in detail the ways in which early learners begin the process of learning to calculate. These two books will expand the research base on which you can draw to develop your practice as well as offering lots of practical approaches to learning and teaching calculation skills.

Thompson, I. (2008) *Teaching and Learning Early Number*. Buckingham: Open University Press.

Self-audit

Carry out this audit to explore how your learning has progressed as a result of working on the ideas in this chapter. Include the results in your portfolio.

1   Make a copy of this table.

Just know it	Need to think a bit	Need to work it out on paper

Using the digits 1, 3, 5, 7, 8, 9 make up a range of addition and subtraction calculations. Make sure you have some calculations that you would place in each column. Write the calculations in the appropriate column together with a rationale for your choice.

2   Set yourself the following tasks:

- an addition involving decimals
- an addition involving decimals you would calculate mentally
- a subtraction involving decimals
- a subtraction involving decimals you would do mentally
- a multiplication involving decimals you could do mentally
- a multiplication involving decimals you would do using pencil and paper
- a division involving decimals you could do mentally
- a division involving decimals you would do using pencil and paper.

Complete the calculations and write a commentary outlining your thought processes and the points at which you think students may make a mistake.

3   Spot the mistake: these three multiplication calculations have got mistakes in them. Write down the mistake that the student has made, then complete the multiplication correctly.

```
    135          270            452
  × 62         × 30           × 18
   270          710        324,016
   710                         452
   980                     324,468
```

CHAPTER 7
PATTERNS AND ALGEBRA

Introduction

Tony tells the story of sitting with a young woman called Natalie who was 14 years old and a student in a class he was visiting to observe a pre-service teacher on placement. He knew that she was not normally in that teaching group. She told him that she had been sent to this group as "the teachers said I was being disruptive." She also said, "I can't do this work. They are doing algebra. I'm in the bottom group so we just do easy stuff like adding and taking away." The problem she was working on was

If you cook a chicken for 20 minutes a pound and then an extra 20 minutes, how long would it take to cook an n pound chicken?

"This is stupid," Natalie told me, "there is no such thing as an *n* pound chicken." Tony asked how long it would take to cook a three-pound chicken. Natalie said, "Well, that's easy. It would be 1 hour 20 minutes because 3 lots of 20 minutes make an hour and another 20 minutes make 1 hour 20 minutes." "So, how about a 5 pound chicken?" Tony asked.

This took a bit longer. Natalie wrote down,

$$20 + 20 + 20 + 20 + 20$$

Then drew a circle around the first three 20s. "That makes 1 hour 40 minutes," she said, "then another 20 minutes is 1 hour 60 minutes, that's the same as 2 hours." "So how do you work it out?' Tony asked. "You count up the 20 minutes, one for every pound and then add another 20 minutes on at the end." Tony was delighted. "You see you can do it," he said.

Natalie looked askance. "Maybe, but there's still no such thing as an *n* pound chicken." she said.

DOI: 10.4324/9781003315155-7

This story represents many people's image of algebra. In the United States the ability to pass SATs in algebra is one of the gatekeepers for moving on in mathematics. In fact, in the US there is a topic called "pre-algebra." We are avoiding this term. There is no such thing as pre-algebra in the same way as there is no such thing as "pre-counting." The moment we start to operate algebraically we are doing algebra. Let's try to demystify the term.

For some, a memory of learning algebra is the moment when mathematics became hard. It is seen as impenetrable. Others wear it as a badge of pride. "Well, I was really good at algebra," they will tell us. All of these responses come from a misunderstanding of what algebra actually is. Many people see algebra as simply substituting letters for numbers and being good at algebra as being able to decode strange sentences such as

$$3x - 7 = 29$$

Algebra is so much more than this. In fact, we are all doing algebra from the moment that we start to notice patterns in numbers and in things like bead strings. Young children are doing algebra when they line up their toy cars on the back of the settee and start to move them around, classifying them according to their own criteria.

This chapter will show you what algebra really is. It will describe how you can introduce algebraic ideas from the moment that children start school and how you can draw on students' intuitive algebraic understandings. It will show you the power that algebra has in generalising patterns that you notice and how it can help you come to a better understanding of the number system. Let us start with an activity.

Starting point

Look at this hundreds chart.

1	2	3	4	5	6	7	8	9	10
11	12	13	14	15	16	17	18	19	20
21	22	23	24	25	26	27	28	29	30
31	32	33	34	35	36	37	38	39	40
41	42	43	44	45	46	47	48	49	50
51	52	53	54	55	56	57	58	59	60
61	62	63	64	65	66	67	68	69	70
71	72	73	74	75	76	77	78	79	80
81	82	83	84	85	86	87	88	89	90
91	92	93	94	95	96	97	98	99	100

Take any 2 × 2 grid in the chart. For example:

12	13
22	23

What do you notice about the numbers in this 2 × 2 grid?

You probably quickly realised that numbers increase by 1 as you move 1 column across the hundreds chart and increase by 10 as you move down the rows. Because of this the number across the diagonal from the number in the top left-hand square is 11 more. Before you read on think if you can calculate the total of the numbers in the 2 × 2 square if you are given the number in the left-hand column. Work on this in your notebook.

Now let us use the information from the paragraph above to generalise, that is to write down the numbers that appear in any 2 × 2 chunk of the hundreds chart:

n	$n + 1$
$n + 10$	$n + 11$

This simply says that if the number in the top left-hand square is n (any number) we can find the number to its right by adding 1, the number below it by adding 10 and the number across the diagonal by adding 11.

We can find the total of all four numbers by adding them:

$$n + n + 1 + n + 10 + n + 11$$

We know that it does not matter what order we add in so we can rearrange this as

$$n + n + n + n + 1 + 10 + 11$$

which is the same as

$$4n + 22$$

So, we can find the total of any 2 × 2 chunk in a hundreds chart by multiplying the number in the top left square by 4 and adding 22. Try it. Try it for several 2 × 2 chunks. See, it always works. We have just used algebra to generalise a result through writing an equation.

Algebra can be described as the ability to understand patterns and relationships. In the previous example we used algebra to describe the patterns in a hundreds chart and to generalise about the relationships between the numbers in the hundreds chart. It is also the ability to use symbols to represent and analyse these patterns and relationships. The moment we moved to generalisation we used the symbol *n* to represent any number.

TAKING IT FURTHER: **FROM THE RESEARCH**

In 2009 a team of researchers including Anne Watson shared the research they had carried out exploring how children learn mathematics. A section of this research discussed the learning and teaching of algebra. They described the nature of school-level algebra. This consists of:

- generalising laws about patterns and number
- modelling mathematical situations
- transforming and solving equations
- manipulating symbolic statements about relationships
- learning how to describe change and relationships using algebraic symbols.

They also suggested that students have natural propensities that can support them in learning algebra. These include:

- interpreting symbols and diagrams in terms of their previous experience of similar symbols and forms
- looking for patterns all around them – very young learners have a natural propensity to classify and sort
- the development of personal mental calculation strategies that they can generalise across similar situations
- the ability to connect, relate to and compare different representations of the same situation.

You may find it useful to reflect on the times that you have noticed the students that you have worked with showing the natural propensities that Anne Watson describes in the aforementioned research.

Progression in patterns and algebra

Foundations for patterns and algebra

Young learners will sort and classify objects that they are familiar with and explain their classifications. They will create, copy and continue patterns. They will be able to tell you what shape or colour should come next in a pattern. This is the beginning of algebraic thinking.

Beginning algebra

Students begin to identify and represent numbers using objects and pictorial representations. They read and write mathematical state-ments involving = signs and can solve missing number problems such as 7 = __ − 9. They come to understand that addition can be carried out in any order but not subtraction and can use the inverse rela-tionship between addition and subtraction to solve missing number problems. They will be beginning to explore number patterns formed by skip counting and be able to describe them.

Becoming confident in algebra

As students become more confident they will identify and write rules for number patterns. They will also be able to use these rules for find missing terms.

Students will represent word problems using number sentences involving all four operations; they will also be able to pose word prob-lems based on numbers sentences they are given. Students will be able to find unknowns in number sentences involving addition and subtraction using partitioning and other strategies.

Extending learning in algebra

Students will continue to create and explore patterns using fractions, decimals and whole numbers and will find unknowns in number sen-tences, which involve multiplication and division, describing the rules used to create the numbers sequences. This will include exploring the numbers of tiles in geometric patterns or creating repeating patterns with shapes (see portfolio task 7.1)

Students will also explore the use of brackets and understand the rules for the order of operations.

They begin to solve scaling problems and can use ratio and propor-tion in simple contexts. They can also work on combination problems. For example, "If I have three hats and four coats how many different outfits can I put together?" As they extend their understanding stu-dents will be able to solve more complex scaling and combination problems such as the number of choices on a menu. They will also begin to generalise mathematical situations such as the example used in the starting point to this chapter. They will begin to express area and perimeter as a formula (perimeter = $2(a + b)$ where a and b are the lengths of the sides). At this stage students will be able to describe and continue number patterns identifying the relationship between each number.

Big idea

One book that has been a great influence the learning and teaching of algebra is *Developing Thinking in Algebra* by John Mason, Alan Graham and Sue Johnson-Wilder. The writing team suggest that:

> *Expressing generality is entirely natural, pleasurable and part of human sense making. Every learner who starts school has already displayed the power to generalise and abstract from particular cases and this is the root of algebra.*

But what does this mean? Maybe three stories from Tony's children beginning to express generality will help. His youngest son, Sam, is now 32 and a lecturer in political science, but at age three he was just starting to make sense of the world. The family were having breakfast one morning and, as Sam ate his cereal, he looked at a photograph sitting on top of the piano. This was a photo of Tony's wife, Helen, as a teenager with her pet beagle, who was called Sam. He glanced down at his cereal, looked back up at the photograph and with that look young children have that suggests great wisdom asked, "How old was I when I was a dog?" In the world of a three-year-old it would make much more sense if everything called Sam was the same thing, rather than having lots of different things called Sam. There is a kind of sense to that and there is also the idea of generality. At a similar time, Sam's sister, Holly was out for a walk with and caught sight of a cat. "Oh look," she said, "a Henry." At that time, she had a dog called Henry and, perhaps, Holly had seen a small furry animal with four legs and a tail and assumed that all these creatures were called Henry. This is generalisation.

Finally, Tony was playing with an old model railway with his grandson Felix, age two and a half at the time. They began to unpack the model trains from their boxes. The first was a "diesel locomotive," and Tony read the label on the box to Felix. This was followed by two "steam locomotives." Felix lined them up on the train track. "Look Grandad, he said, "all locomotives." This is what John Mason is describing above.

So – the big idea in algebra is . . .

Generalisation

Professor Jo Boaler from Stanford University has put together a MOOC (massive open online course) called *How To Learn Math*, and several of the videos (available on YouTube) explore the learning and teaching of algebra with a focus on generalisation. Jo argues that the current teaching of algebra is an equity issue as the proportion of learners who fail examinations due to their lack of confidence in algebra come disproportionately from low-income families. She argues that the current teaching of algebra focuses on procedural algebra (finding a single unknown number) rather than structural algebra where we use symbols to express generality. You will notice that we chose to start this chapter with a piece of structural algebra. You may also notice that curriculum documents tend to focus on procedural algebra such as finding a missing number, which is unfortunate. Try this portfolio task as a way of seeing what we mean by structural algebra. It is a task you may recognise from your time in school and is often used in the later stages of secondary school. I would argue tasks can be worked on with learners in primary school.

Portfolio task 7.1

Look at this series of shapes:

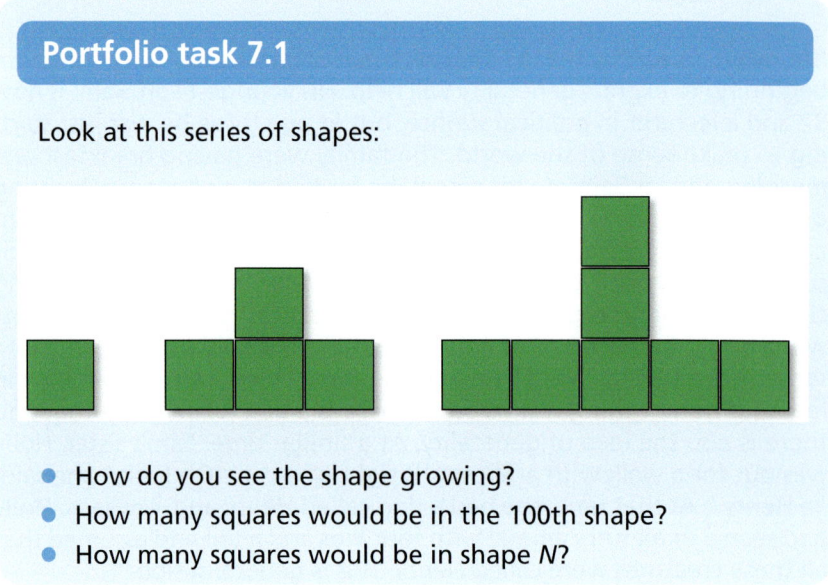

- How do you see the shape growing?
- How many squares would be in the 100th shape?
- How many squares would be in shape N?

If you do not know where to start on this get yourself some cubes and begin to build the shapes. What would be the next shape in the sequence? Write down everything that you notice about the shapes in your notebook. Imagine that you are trying to give someone else instructions on how to create the sequence.

If you worked on this in secondary school, you were probably asked to complete a table such as this one rather than exploring for yourself.

Shape number	Squares
1	1
2	4
3	7
4	10

You then will have noticed that the number of squares was increasing by 3 each time. This is called a linear relationship (if we were to draw a graph of the results, we would get a straight line – try it). You will also have realised that this did not really help you find how many squares were in shape 100 or in the nth shape as all you could do was add 3 on to the previous shape. This comes from jumping straight to procedural algebra.

Rather than this, think about the first question, "How do you see the shape growing?" When working with primary students on activities such as this they do not seem to feel bound by having to jump straight to an equation and so they will say things like,

In the third shape there are 3 legs that are 2 squares long and a square in the middle. In the 4th shape there are 4 legs that are 3 squares long and a square in the middle.

Here they are expressing generality. For any shape the length of the leg is one less than the shape number. For the 100th shape the legs would be squares 99 long. And then there is the square in the middle. So, we can find out how many squares by calculating:

$$99 + 99 + 99 + 1 \text{ or } (3 \times 99) + 1 = 298$$

If I want to describe the nth shape each leg would be $n - 1$ squares long, and then there is the square in the middle, so I can write

$$3(n - 1) + 1$$

And this formula works. Try it for the numbers in the table. This is how we can begin with structural algebra and focus on generalisation rather than only allowing students to see algebra as a procedure in which we have to find missing numbers or in which letters stand for numbers that we have to find.

TAKING IT FURTHER: **FROM THE CLASSROOM**

In a paper called "Algebra in the Early Years: Yes" published in 2003 in the US journal *Young Children*, Jennifer Taylor Cox describes a wide range of activities that she engages in with very young learners to support them in developing early algebraic thinking in the first stages of schooling. She describes what algebra in the Early Years looks like, showing how young children can describe position patterns using vocabulary such as "up, down; up, down" or "top, side, front, top, side, front"

You will recognise this use of language from the young learners that you work with or may be from your own young children or nephews and nieces that you play with. Teachers can build on this intuitive use of language by looking at other attributes such as shape or colour. She also shows that young children are able to build growing patterns using cubes following patterns that teachers set up or by making their own growing patterns.

She uses pan balance scales to introduce the idea of equality. If things balance, then they are equal. The children also realise that sometimes things need to be added to one side of the balance to make it equal.

Finally, she shows that using pictures such as images of cats can support us in asking questions such as, "if we have four cats how many ears are there? How many paws?" This is another early algebraic activity.

Whenever this article is used with Early Years educators they are convinced that they can begin to introduce algebraic thinking into the activities they provide. In fact, the most common response is, "we have been doing algebra for years and we didn't realise it."

The following teaching points suggest ways that you can introduce algebra to students in your classrooms and develop all students' capacity for algebraic thinking.

Using Cuisenaire rods to develop algebraic thinking

You might assume that algebra is an abstract idea and so cannot be modelled using concrete materials. Far from it. In the previous chapter we showed you how young learners automatically used algebra to describe the relationships between a pattern of Cuisenaire rods. In the Association of Teachers of Mathematics book, *Cuisenaire – from Early Years to Adult*, Simon Gregg, Mike Ollerton and Helen

Williams illustrate how the rods can be used right across the mathematical world.

The rods can be used to build growing patterns.

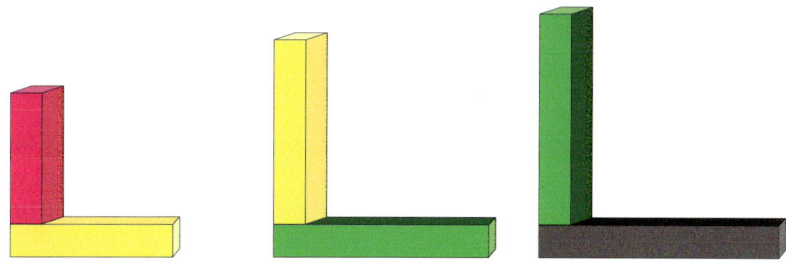

I would be asking students to think about what the first step in the pattern would be and to explain how it grows.

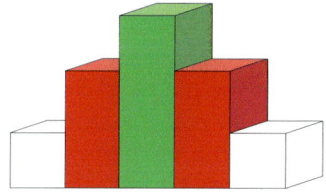

They can use the rods to create staircases, which they discuss in the same way. Simon, Helen and Mike also show how the rods can explore the relationship between square numbers and sums of consecutive odd numbers

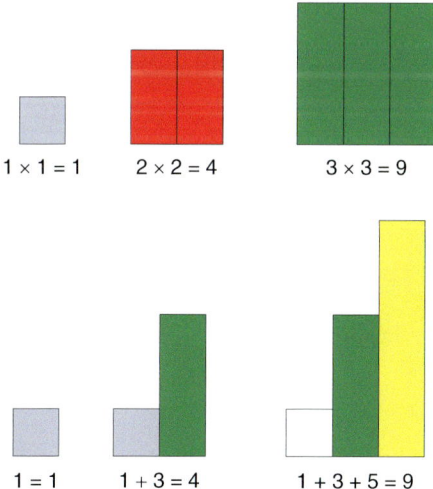

We hope to have convinced you of two things. Algebra is accessible to all learners and Cuisenaire rods are indispensable in the mathematics classroom.

Teaching points

Teaching point 1: describing and creating patterns

Young learners will sort and classify objects intuitively. As John Mason pointed out earlier in the chapter many children come to school with this as a behaviour that they have developed at home. It is helpful to join in young children's play and try to continue their patterns.

It is also helpful to join in the play and create your own bead strings, talking about the pattern as you go along. "I'm putting two blue, then two red, then two green and then another blue. What should I put next?" Encourage students to talk about the patterns that they make and to work with each other to copy and re-create patterns. You can develop this, using shapes that repeat, colours, or even a mixture of shapes and colours.

It is important to show the students that the pattern repeats, so you have to see the repeat rather than just show them the first part of the pattern. As students move through school, they can move on from this to look at the sort of growing patterns that you saw earlier in the chapter. For example, they could describe how this pattern grows.

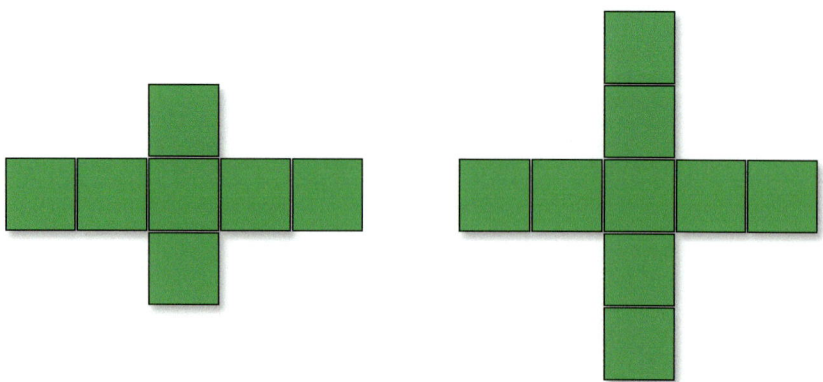

They could create the pattern using linking cubes or rods to explore the pattern physically. Sometime constructing a pattern concretely can help a student notice the pattern they are creating.

Teaching point 2: developing an understanding of the "equals" sign and equations as "balanced"

As suggested earlier, balance scales are a great way to begin this process. Students can use a range of objects to balance against each other. They could see how many cubes will balance a toy bus and begin to record this using pictorial representation. Here they might actually draw the scales with images of the bus and the cubes on each balance pan. Some will use iconic representation. They may draw lines for the cubes on one scale and a box for the bus on the other scale.

Some students may even write B = 4C

This type of activity supports students when they meet the missing number problem such as

$$8 - \boxed{} = 2$$

as they have an understanding of the importance of "equality." Activities that allow students to see that symbols can "stand for" a range of different numbers are also important.

$$\bigcirc + 8 = 10n + \triangle$$

Teaching point 3: modelling quantitative relationship

As you will realise by now, Tony has a young grandson called Felix. He is just approaching his ninth birthday as we write this Australian edition of the book (something he is very excited about). He still spends a lot of his time playing with toy cars and trucks and is also using Lego creatively. From the age of three he has been classifying then in many different ways and will explain the criteria by which he is sorting them in great detail. He then uses these classifications as the basis for complex narratives that he builds around the cars and trucks. He uses properties of Lego bricks to describe, classify and sort them.

Felix started to "model quantitative relationships" from the moment he could count, by exploring how many wheels there are on different numbers of cars. The family has four dogs offering lots of opportunity for counting ears, eyes and paws. This is the beginning of seeing the relationships between quantities.

Try this portfolio activity to see how this type of activity can be developed.

Portfolio task 7.2

A farmer has only goats and chickens. She has 28 animals and there are 72 legs altogether.

How many chickens and goats are there?

You may find it helpful to model the animals.

Teaching point 4: describing change

Quantitative changes are an everyday part of children's lives. They realise that their shoe size changes as they get older. They may well have a measure on the wall at home which shows how their height changes over a period of time. You can draw on these experiences at school. You can also set up activities involving change such as growing sunflowers and measuring the rate at which the height changes. Similarly, science experiments such as exploring how much further a toy car will travel if you increase the slope of a ramp introduce students to the notion of describing and recording change. The following portfolio task requires several friends. This is one of the few activities Tony has carried out in a bar! He had used the activity with a friend in her classroom during the day and she was so excited that she made him repeat it in the bar that evening, everyone in the bar joined in.

Portfolio task 7.3

What do you think the ratio of the circumference of your head to your height is?

Make an estimate and then collect the information from lots of your friends.

You could construct a scatter diagram to work out the relationship.

Do you think the ratio would be the same for the children that you teach?

What about the children in the local kindergarten?

In practice

The following lesson plan was used with a group of Year 3 students. They had not been prepared for the activity but carried it out just before their Christmas party, so it became part of their party planning. Following the plan is an evaluation of the lesson that explores how successful the plan was in supporting the children to develop their knowledge, skills and understanding.

Learning intentions

- develop and use written methods to record, support and explain multiplication of two-digit numbers by one-digit numbers
- some students will begin to generalise the solution.

Key vocabulary

number patterns
Resources: Rolled up paper to model crackers, mini whiteboards to draft ideas on. Large pieces of paper to represent the problem.

Context for lesson

The students are planning a Christmas party and wanted to have crackers to pull as a part of the celebration. They had asked how many crackers I thought they should buy. This activity was to help them come to a decision.

Warm-up activity: what's next

The following sequence of numbers was placed up on the whiteboard:

1, 3, __

Students were then asked which number might come next. Henry put his hand up first and suggested the next number was 5, which was then recorded on the board. I then asked, "What else could come next?" At first, very few hands went up, but after providing some wait time, there were many students to pick from. As they gave me other alternatives, each one was recorded on the board. In the end, we had six possible answers on the board:

5, 6, 9, 7, 13, 2.

Students were then asked to pair up and find out what the rule for each possible answer was. They were given about three to four minutes to complete this task and were asked to record their thinking in their books. This was then followed by a whole class discussion, where groups shared their rules. Chloe went first and she said that the rule for 1, 3, 5 was easy, as "you just add 2 each time." Next, Sam told the class that the rule for 1, 3, 6 was that "the number you add on is getting bigger by one each time." When I asked her to explain this further, she said, "Well, the first time, you added a 2, so it went from 1 to 3. The next time you add a 3, so it goes from 3 to 6. And the pattern just keeps on going." I asked the class what number would come after 6 if we followed Sam's rule, and Mikayla correctly identified that it would be 10.

The final sequence of numbers that we discussed as whole class was 1, 3, 7. Max put his hand up and said that he thought it was "double the previous number and then add one more." I said, "So the next number would be 15." And Max said, "Yeah, because double 7 is 14, then add one more is 15."

Launch

The question of how many crackers we should buy was posed to the students. After a brief discussion, we came up with the shared understanding that it would be good if each student could pull a cracker with every other child.

Explore

Students approached the task is variety of ways. Some used rolled up paper as crackers, some drew diagrams, some tried to organise their thoughts into a table.

There were a few students who were finding it difficult to get started. In these instances, I suggested they try the simple cases first and model it physically. So how many crackers for one person, for two people and so on.

Summary

The class was brought back together and selected groups were asked to share what they found out. The emphasis of this discussion was the processed used and the learning that took place, rather than the solution.

Rationale and evaluation

It was great to be able to use a question that had come from the group themselves to introduce this investigation. As we had begun to discuss the question in a previous lesson the "talk partners" at the opening of the lesson proved an effective way of generating lots of different answers. Some pairs started off by thinking it would just be one cracker each so gave the answer 28 (the number of students in the class). Amina quickly countered this: she said it would be loads more than that because you will use up your cracker straight away on the first person and there would still be 26 people left to for you to pull a cracker with. This seemed a good point to set the groups off on the investigation.

We worked in our collaboration groups and spent some time discussing the different roles people would take. The groups all chose someone to "lead" the group. It was their job to make sure that everyone was listened to. The groups all chose someone else to keep notes of the discussions. I had been concerned that some of the groups might not get started but Rupna's group immediately used the paper to make pretend crackers and then stood in a circle to count how many crackers they would need. Most of the other groups followed this lead as they saw what they were doing.

Simon's group operated slightly differently and wrote their names down on a piece of paper and drew lines to connect each name to every other name. They quickly came to an answer for their small group (of five) but were struggling to think about generalising this.

At this point I drew the whole class back together and asked Rupna and Simon to share their approach to the problem. This led to some really useful discussion about how best to work on an answer for the whole class. No one had thought about starting with the simplest case and working up from that. I suppose this isn't a surprise as it's not necessarily something that the students are used to. So I modelled this with the whole class. We agreed that if there was 1 person there would be no crackers, and for 2 people it would be 1 cracker (at this stage Amina immediately shouted out, "it's just one less"). Then we looked at 3 people and this gave us 3 crackers. Amina realised she wasn't correct. Some of the groups had already worked out the answer for 4 and some for 5, so I asked all the groups to go back and explore the answers for groups of 6 and 7 people. I didn't specify that they should put the results in a table. Some groups used a table to collate the results whilst some used a circle to illustrate the answers with the total underneath. For example, this is the diagram for 6 people.

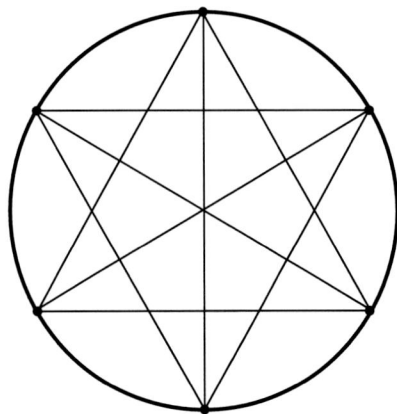

I pointed out that this way of illustrating the problem is called the **mystic rose**. It also quickly shows that the answer for 6 people is 6 × 5 = 30.

We all came back together again and shared our learning. I wanted the focus to be on learning rather than simply the answer, although Amina was very quick to point out her "mystic rose" and how it had given them the answer very quickly.

Finally, we decided it would be too expensive to buy 276 crackers so we would make do with 1 cracker each.

Portfolio task 7.4

Devise a lesson plan that is appropriate for a group of students that you are working with.

The focus should be an aspect of algebra. You could choose from some of the ideas in the "Teaching points" section of this chapter or find another appropriate activity. Think carefully about the students' previous experience to make sure that the activity is accessible and challenging. Once you have taught the lesson evaluate it carefully and add this evaluation to your portfolio too.

Assessing algebra

The focus on assessing algebra should be practical tasks which move from specialising to generalising. These will take a different form depending on the age of the students that you are working with, however, make sure that the tasks are sufficiently open so that you can observe and question the students to elicit their current level of understanding and move them forward in their thinking.

For the youngest learners focus on making repeating patterns with coloured beads and with shapes. You can use the sand and pastry cutters or different styles of potato prints. You could even make very simple repeating patterns with the numbers of cherries on cakes. You can also see which criteria students are using to classify objects. Developing language such as *bigger* and *smaller*, *shorter* and *taller* is important for the development of algebraic ideas. You could also use the balance scales as discussed earlier in the chapter.

As students develop their understanding, focus on the structural aspects of algebra rather than the functional aspects in your assessment of algebraic thinking. So, whilst the missing number problems can help you see how well students can remember number facts at this stage, they do not really help you support algebraic thinking. Instead, develop the types of activity discussed earlier by forming patterns of cubes or squares. Use the activities introduced earlier in the chapter that allow you to move from specialisation to generalisation. This is the big idea we discussed earlier. This is the skill that your assessment should focus on.

Cross-curricular teaching of algebra

The idea of the extended project is to draw on the key ideas within the chapter to develop cross-curricular activities which you can explore with your learners over a series of lessons. This allows you and your class to develop your subject knowledge together. The previous lesson plan shows you one possible activity built around a celebration.

Another activity that you could develop could be constructed around a play or other entertainment at the festive party. For example, you could devise a play that includes butterflies (with two wings); sheep (with four legs); and spiders (with eight legs) and plan combination problems using these facts.

If it is winter celebration you can construct many different patterns using a snowflake as an inspiration. This allows the students to predict how the pattern would grow and to explore generalisation within the growing pattern.

Summary

The key focus for this chapter has been the all-pervasive nature of algebra within mathematics. You have been shown how algebraic thinking can be developed from the moment that a student enters school and that young children have a natural aptitude towards classifying and then making generalisations based on their observations of the world around them. This natural aptitude can be drawn on in school to develop their algebraic thinking further. The progression section illustrated how these ideas are developed as students move through school.

The big idea is specialising and then generalising, noticing what is particular about a pattern and then being able to generalise this so that we can continue to build the pattern. There were examples used from a paper that illustrated how Early Years teachers can use algebra in a very practical way, and these were developed into the series of teaching points. These teaching points all offered starting points for your own algebraic activity in your classroom.

Reflections on this chapter

Our hope is that this chapter has demystified algebra for you. If you are one of those people who loves algebra and remembers feeling a great sense of achievement as you solved complex equations, we hope that you can see how you can begin to share this excitement with young students. They do not have to wait until secondary school to engage with algebraic ideas and to experience the delight of being able to explain how a pattern is growing and explain the generalisation that they are seeing. Or you may be one of those people who has always been scared of algebra, who remember it as a puzzle that you could never solve or a strange language that you did not have access to. We hope that this chapter has shown you how you can explore algebraic ideas with your leaners and so become algebraic thinkers together. Try the activities for yourself, complete the portfolio tasks and then enjoy working with the learners in your class.

Going further

Reasoning in number and algebra

LORRAINE DAY

This article published by the Australian Association of Mathematics Teachers shares a series of lessons which are designed to develop mathematical reasoning skills by connecting reasoning in number and algebraic thinking.

Day, L. (2014) 'Reasoning in number and algebra', *Australian Primary Classroom*, 19(3), 16–19.

Cuisenaire – from Early Years to adult

SIMON GREGG, MIKE OLLERTON AND HELEN WILLIAMS

This ATM book shares many practical classroom activities drawing on Cuisenaire rods. I would argue that all the activities presented here develop algebraic thinking but the whole of the third chapter has a specific focus on *Counting, sequences, patterns and algebraic reasoning*. There are lots of examples of student responses, which makes the activities accessible to all learners and teachers.

Gregg, S., Ollerton, M. and Williams, H. (2017) *Cuisenaire – from Early Years to Adult*. Derby: Association of Teachers of Mathematics.

Developing thinking in algebra

JOHN MASON, ALAN GRAHAM AND SUE JOHNSON-WILDER

This is one of a series of three books written by the team at the Centre for Mathematics Education at the Open University in the UK. The series was written for teachers both to support them in teaching and to help them develop their own mathematical thinking. The books integrate mathematics and pedagogy in the way we have attempted to do in this book. If you only buy one book on algebra, make it this one!

Mason, J., Graham, A. and Johnson-Wilder, S. (2005) *Developing Thinking in Algebra*. London: Paul Chapman Publishing.

'What's *x* got to do with it?'

ANNE WATSON

This article appears on the Nrich website. In the article Anne aims to outline the aspects of algebra that should be introduced to young children if they are going to adopt an algebraic approach to their mathematics. Anne explores the "threads" of algebraic thinking from primary into secondary education. There are also links to a much wider research project into children's mathematical thinking published by the Nuffield Foundation. Watson, A. *What's × Got to Do with It?* Available at http://nrich.maths.org

Self-audit

Look back to the first portfolio task. If you remember, you were asked to explore 2 × 2 chunks of the grid like:

12	13
22	23

1	2	3	4	5	6	7	8	9	10
11	12	13	14	15	16	17	18	19	20
21	22	23	24	25	26	27	28	29	30
31	32	33	34	35	36	37	38	39	40
41	42	43	44	45	46	47	48	49	50
51	52	53	54	55	56	57	58	59	60
61	62	63	64	65	66	67	68	69	70
71	72	73	74	75	76	77	78	79	80
81	82	83	84	85	86	87	88	89	90
91	92	93	94	95	96	97	98	99	100

Explore shapes like this within the grid. Can you find a generalisation to give you the total if you know the number in the middle square?

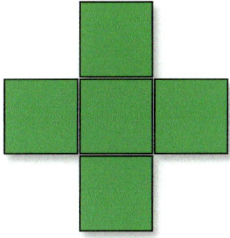

What about a cross with 4 squares in each leg, or 5?

What about shapes like this?

CHAPTER 8

GEOMETRY
Properties of shapes; position and direction

Introduction

Tony remembers working in a mixed age class in a small school in the north of England. There were children aged from 7 to 11 in the class and the teacher was working on an activity involving visualising whether or not a range of **nets** of cubes would "fold up" to make a cube. The students had created the nets by drawing all the different arrangements of six squares they could think of. Once they had decided which they thought would fold up into cubes they cut them out to check their conjectures.

At the end of the session the teacher said to Tony, "What was interesting to me was that some of the less able children did best on that activity." You may well have heard similar unthinking use of the word "ability." We would suggest that if someone has been successful in a mathematical activity linked to shape, surely they are "able" even if they are challenged in other areas of mathematics are not at similar levels. Some of you will be more comfortable with this area of mathematics; for some of you it will present a challenge. We hope that this simply illustrates the breadth of the mathematics curriculum and how there will be areas of mathematics in which we feel comfortable and areas in which we struggle. The important point is that we are able to make connections between all areas of mathematics so that we become confident mathematics teachers and learners.

Starting point

Close your eyes. That may seem a strange request to someone reading a book, so let us expand. First, find a friend, then ask them to read the

DOI: 10.4324/9781003315155-8

next section whilst you have your eyes closed. If we are going to think about how you teach and learn geometry we need to begin with a visualisation.

> Create a large red rectangle that you can see clearly in your mind's eye.
>
> Stand it on its end so that you can see it standing on its shortest side.
>
> Then slowly rotate it so that it is laying on one of the longest sides.
>
> Now rotate it again so that it is balancing on a corner.
>
> Move the rectangle round and round in your mind's eye and make a decision about which way round you want finally to picture the rectangle and stop it in that orientation.
>
> Now imagine a small, blue right-angled triangle that will fit inside the rectangle.
>
> Picture it inside the rectangle and slide it so that the right angle fits exactly into one corner of the rectangle.
>
> Imagine another small, blue right-angled triangle, which can be a different size from the first one you thought of.
>
> Slide it into one of the other corners of the rectangle. Notice the red shape that is left inside the rectangle.
>
> Open your eyes and sketch what you see.

Talk to your friend about the shape you have sketched. How many different properties can you describe? What do you notice about its properties? Are there parallel lines (like railway lines) or perpendicular lines (lines at 90° to each other)? What can you say about the angles: are they acute (less than 90°), obtuse (between 90° and 180°) or reflex (more than 180°) angles? Do you know what the shape is called?

This simple activity embraces two of the key skills you need to teach students to help them understand shape. They need to be able to visualise the shapes that we are working with and they need to be able to describe them so that someone else can visualise the same shape. We do this by having a clear understanding of the properties of different shapes.

These are both skills that we can teach, although visualisation is not always a skill that we will have been taught. As we suggested in the opening of the chapter, the interesting thing is that sometimes those students who have not excelled when working with numbers will be able to visualise and manipulate shapes very quickly. These students are good at mathematics too.

Portfolio task 8.1

Look at the polygons below:

Classify them in any way you like, into as many groups as you like. When you have completed this task explain your classification to a friend. Then classify them in a different way.

Finally, name as many of the polygons as you can.

(This activity is taken from many of the free online resources provided by the National Council of Teachers of Mathematics in the USA.

This activity can be found at http://illuminations.nctm.org/LessonDetail.aspx?ID=L277)

See end of chapter for possible solutions.

You will have noticed that you are having to draw on language that you may not have used for a while. There are the names of the shapes (all provided in the answers at the end of this chapter) and perhaps, more importantly, ways to describe their properties. We have already used **parallel** and **perpendicular**. Other vocabulary you may have drawn on for the previous activity is **congruent shapes** (two shapes which will fit perfectly on top of each other) and **similar shapes** (these are shapes whose sides and angles are all in the same ratio).

For example, the two right-angled triangles in Figure 8.2 are similar as one has each side three times bigger than the other one. This also means that the **corresponding angles** in each triangle are the same size.

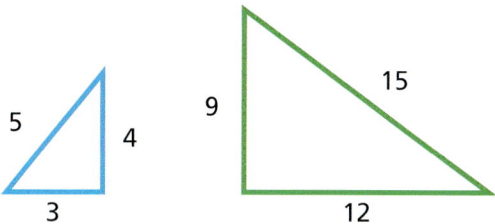

Figure 8.2

Taking care to develop student's vocabulary of shape is vital, so make sure there are plenty of displays utilising the vocabulary and also that there is plenty of opportunity to talk about the shapes we are working with.

In the 1950s Pierre van Hiele and Dina van Hiele-Geldof, two Dutch mathematics teachers and husband and wife, identified five levels of understanding that help us appreciate how students come to understand shape. These are described in Pierre's 1999 article "Developing geometric thinking through activities that begin with play" in *Teaching Children Mathematics*. It is useful for us to explore these levels in some detail before exploring how progression is described by the strategy:

Level 0:   **Visualisation** – at this stage students can name and recognise shapes by their appearance. They cannot yet describe properties or use properties to sort shapes.

Level 1:   **Analysis** – at this stage students identify properties related to shapes and use these to classify them.

Level 2:   **Informal deduction** – at this stage students can use the properties that belong to classes of shapes to problem-solve. So, they will be able to talk about **regular** and **irregular shapes** and about triangles in general or specific types of triangles. This is the furthest we would expect most learners to progress within primary school.

Level 3:   **Deduction** – at this stage students use their understandings about shape to construct geometrical proofs. That means that they can use their understanding and knowledge about the properties of shapes to convince others that new understandings are true.

Level 4:   **Rigour** – at this final stage students would be constructing rigorous proofs about the geometrical properties of shapes. These proofs will follow mathematical conventions rather than the more informal proofs at level 3.

This research suggests that the visualisation and analysis stages of learning are key to understanding shape. This has become the focus of most of the teaching throughout the early and primary years of education. The teaching points later in the chapter emphasise the importance of the development of the vocabulary to allow students to be able to talk about what they are noticing at the visualisation stage. You will notice that many students will move between the visualisation level and analysis level as their vocabulary develops to describe particular concepts within understanding shape. As with many "schemas" for learning, this can be seen as dynamic rather than a stepladder. That is, students do not rigidly move through the four stages; rather, they will move backwards and forwards through the stages depending on the particular idea they are exploring.

TAKING IT FURTHER: **FROM THE RESEARCH**

Penny Coltman, Dinara Petyaeva and Julia Anghileri have explored how best adults can support young children using building blocks to carry out problem-solving activities that help them come to an understanding of three-dimensional (3D) shape. In an article "Scaffolding learning through meaningful tasks and adult interaction," published in the *Early Years Journal*, in 2002, they suggest that young children between three and six, although operating at the first of the van Hiele levels, are limited by their experience and language and that through careful adult intervention children can learn effectively.

The children were given *poleidoblocks*, a set of wooden 3D shapes, and toy animals and cardboard models to produce playful contexts that made sense to the young children. Children were allowed to become familiar with the blocks through free play and constructed their own stories using the blocks and the other resources. However, after this free play the researchers then used practical activities to support the children's learning.

For example, children were introduced to cylinders and cuboids. The children were told a story about cylinder birds loving to roll – the children could use this idea to see the difference between cylinders and cuboids and to begin to sort them according to this property of "rolling."

This research emphasises the importance of exploring properties of shapes by manipulating the shapes themselves. Shapes are dynamic; they move in space: 2D shapes move so that we can see them in any orientation; 3D shapes can roll and slide and build and balance. By working practically with shape, children are much more able to "see" their properties.

The next section outlines the development of children's understanding about shape as they move between foundation stage and the end of their time in primary education. You should keep in mind the advice from the aforementioned research that as much of this as possible should be explored actively and dynamically.

Progression in geometry

Foundations for geometry

The initial focus should be to provide children with common objects and shapes so that they can sort and describe them, build models and patterns using 2D and 3D shapes and talk about what they are doing to begin to develop the language of shape, size and position. Also have conversations with the students about the shapes they see in the environment. They will begin to classify shapes that they recognise. By the end of this stage students will be describing and drawing 2D shapes and describing 3D shapes using a range of properties.

They will also be able to give and follow directions to places they know. This will develop so that they can interpret simple maps and describe relative positions of places on the maps.

Students should be introduced to ideas of whole, half and quarter turns as an introduction to angle. They can be introduced to the difference between things that turn about a point (like a pair of scissors)

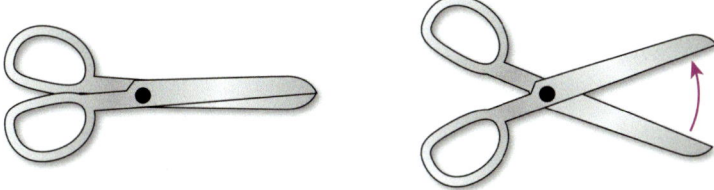

or about a line (like a door). You can help them develop their use of language to describe position, using vocabulary like in front of, behind, next to, on top of and underneath.

Later you can teach students to investigate the effect of one-step slides and flips.

Becoming confident in geometry

At this stage students will be making models of 3D shapes and describing them. They will be able to compare and create **compound shapes** (shapes made by combining two or more single shapes).

Students will be able to create maps on grids, which they will use to describe the position of objects on the grid. They will use simple scales, legends and direction to interpret these maps.

Students will be able to compare amounts of turn using angles. They will also be able to identify symmetry in the environment. Later they will create symmetrical patterns, pictures and shapes, sometimes using digital technologies.

Extending learning in geometry

Students at this stage will be able to connect 3D objects with their nets and with other two-dimensional representations. They will use this skill to help them construct simple prisms and pyramids.

They will use a grid reference system to describe routes and locations leading to introducing the **cartesian coordinate system** in all four **quadrants**. This will help them in describing and investigating **enlargements**, **translations**, **reflections** and **rotations**.

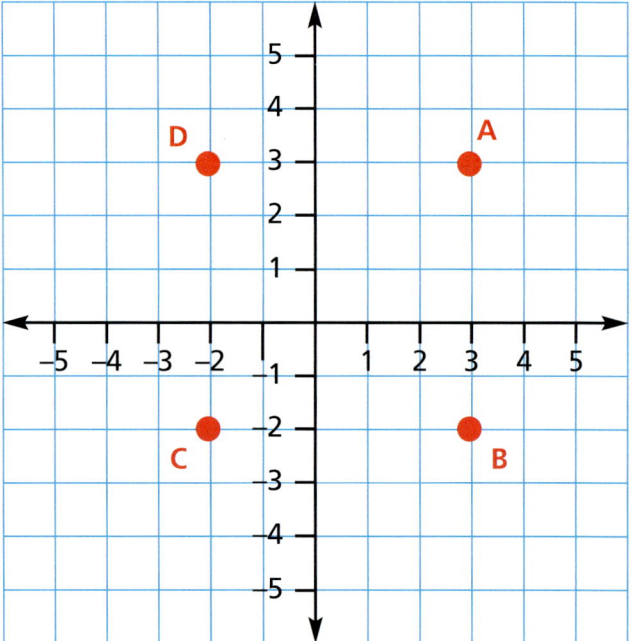

Finally, students will estimate, measure and compare angles using degrees and will be able to use a **protractor** to measure and construct angles. They will investigate angles on a straight line, angles at a point and **vertically opposite angles** (see following diagram).

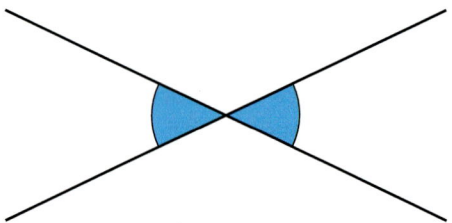

Big ideas

Properties of shapes

A student in an algebra lesson said to his teacher, "That's what maths is all about, isn't it? It's about saying complicated things very simply." Being able to classify and order sets of objects fits this definition perfectly. We classify and order shapes using their properties and we need to be able to explain the similarities and differences as simply as possible. A property of a shape is something that does not change. It might be 2-dimensional (2D; having length and width but not depth) or 3-dimensional (3D; having length, width and depth); shapes can be **closed** (all sides join together) or **open** (there is a gap between two corners); they may have sides that are all **straight** or some **curved** sides; shapes can be **convex** (all the interior angles are less than 180°) or **concave** (at least one interior angle is greater than 180°) or they may be regular (all sides the same length and all interior angles the same size) or irregular (having sides and angles that are different lengths and sizes).

   Closed shapes with straight sides are called polygons, and these are listed in Figure 8.6.

Number of sides	Names	Regular	Irregular
3	Triangle		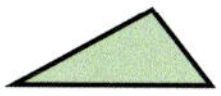
4	Quadrilateral		
5	Pentagon		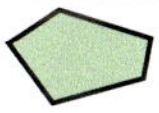
6	Hexagon		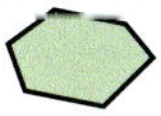
7	Heptagon	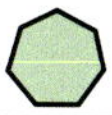	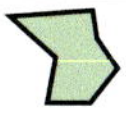
8	Octagon	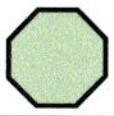	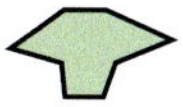
9	Nonagon	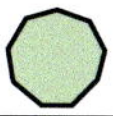	
10	Decagon	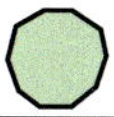	
11	Hendecagon	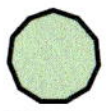	
12	Dodecagon	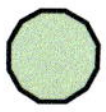	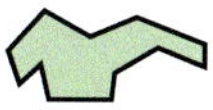

Figure 8.6

It is important to use both regular and irregular versions of the polygons with children, otherwise they only link the name to the regular version of the shape.

Within the set of polygons there are further classifications. So, for example, triangles can be:

Equilateral   All angles are 60° and all the sides are the same length.

Right angled   One angle is 90°.

Isosceles   One pair of sides are the same length.

Scalene   All sides are different lengths.

A triangle can be right angled and isosceles or right angled and scalene.

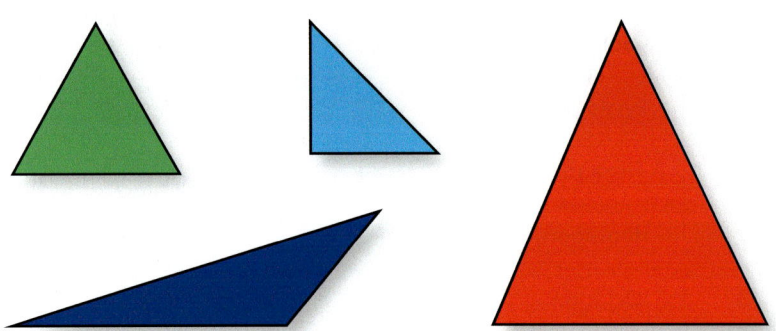

Similarly, several quadrilaterals have special names:

Square   A quadrilateral with four equal sides and four right angles.

Rhombus   A quadrilateral with four equal sides; the angles do not all have to be 90° (so a square is a specific version of a rhombus).

Rectangle   Any quadrilateral with four right angles. An **oblong** is a rectangle that isn't a square.

Parallelogram A quadrilateral with opposite sides parallel.

Trapezium   A quadrilateral with only one pair of parallel sides.

Kite   A quadrilateral with adjacent sides (i.e. sides that are joined at a point) of equal length.

Portfolio task 8.3

Try this activity – the folding helps you understand ideas of symmetry and the surprises when you unfold the shape support you in noticing the properties of the shapes. Students find it motivating too as it develops their skills of visualisation as they have to try to visualise the shape that will emerge when they unfold the paper.

Take a piece of A4 paper and make one fold anywhere. You do not have to fold it in half. Make a cut so that you form two shapes out of the piece of paper. Sketch the two shapes that you have made and name them. Can you find ways to fold and cut the paper so that you make a square, a rhombus, a rectangle, a parallelogram, a trapezium and a kite?

3D shapes can also be classified in several ways. **Prisms** and **pyramids** are sometimes confused. A student once described the difference as, "A prism is where you find naughty people and a pyramid is where you find dead people." The actual difference is to do with the cross-section. You can slice a prism at any point, parallel to the face at the end, and you will always get the same cross-section, whereas if you slice a pyramid you will have different sizes of the same shape.

Regular triangular pyramid

Regular square pyramid

Regular hexagonal pyramid

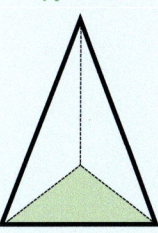
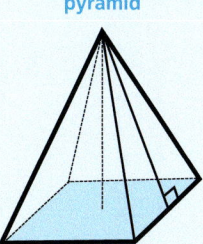
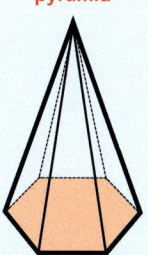

Two prisms you might recognise are a **cube** and a **rectangular prism** The most common 3D shapes are shown in Figure 8.9. One way of classifying 3D shapes is by the numbers of **faces**, **edges** and vertices (singular **vertex**). The faces of a shape are the flat regions. The edges are where two faces meet, and a vertex is a point at which two or more edges meet.

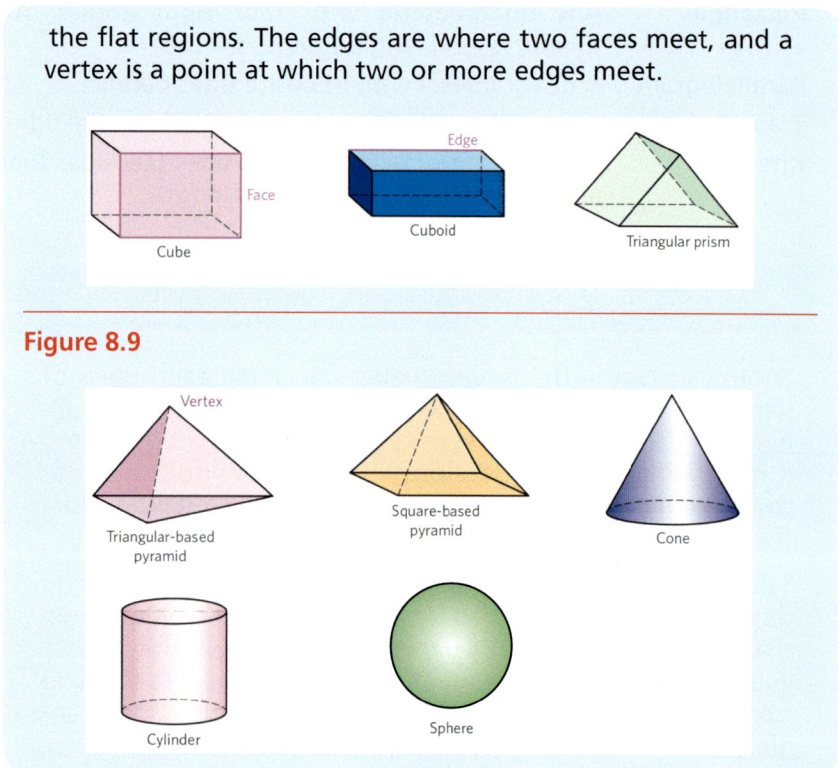

Figure 8.9

Position

When we want to define, or locate, a point mathematically we use coordinates on a pair of axes (you pronounce this "axees," not like the implement you chop wood with). These **axes** cross each other at what is known as the **origin**, which has **coordinates** (0,0). The four areas created by the axes are known as the four **quadrants**. The best way to explain these terms is by a diagram. In the following diagram the **x-axis** is the horizontal axis and the **y-axis** is the vertical axis:

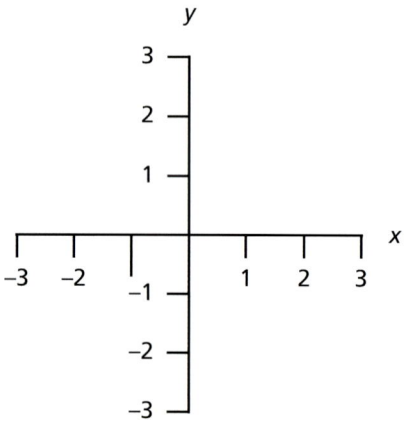

The two axes are labelled x and y. We write coordinates as a pair of numbers in brackets separated by a comma, for example (2,3) or (21,4). The first number always refers to the x-coordinate and the second number refers to the y-coordinate.

Primary students are also introduced to defining direction by compass points.

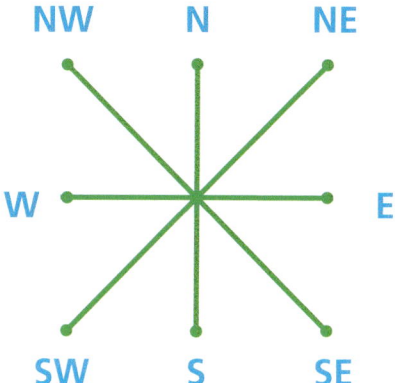

There are four points of the compass: north, south, east and west. The direction halfway between north and west is described as north-west. Similarly, the direction halfway between south and east is described as south-east, the direction halfway between south and west is south-west and the direction halfway between north and east is north-east.

Once we have placed a shape on a grid, we can change its shape through a **transformation**: this is the process of moving a shape on a grid. The three transformations are translation, reflection and rotation.

A **translation** leaves the shape's dimensions and its orientation unchanged. In other words, it is the same as sliding the same shape across the grid. Here is an example from a Year 6 girl:

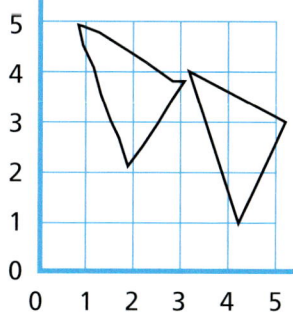

This shape has moved 2 in the x direction (that is, backwards along the x-axis) and +1 in the y direction (that is, upwards along the y axis) and so you would write

$$\begin{pmatrix} -2 \\ 1 \end{pmatrix}$$

A negative sign in front of the number would represent a move backwards along the x-axis or down the y-axis.

A **reflection** transforms the shape using a **mirror line**, as in Figure 8.14:

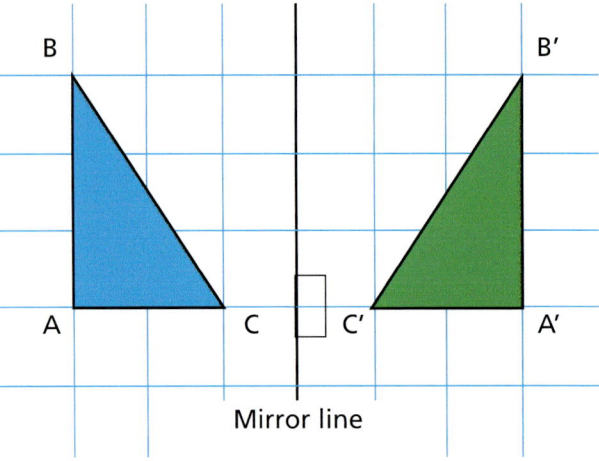

Mirror line

Figure 8.14

When reflecting in a mirror line you need to take care to check that all points are the same distance from the mirror line. So, in the previous example, points A and B are three squares from the mirror line, and C is one square away. You must always measure at right angles to the mirror line. Finally, we can **rotate** shapes.

When we rotate shapes we have to define the point we are rotating about, called the **centre of rotation**, and say the angle we are rotating through. For example, if we rotate this triangle about the origin the dimensions of the triangle will stay exactly the same. It is as if we have fixed a "stick" to the corner "A" and moved it through x degrees:

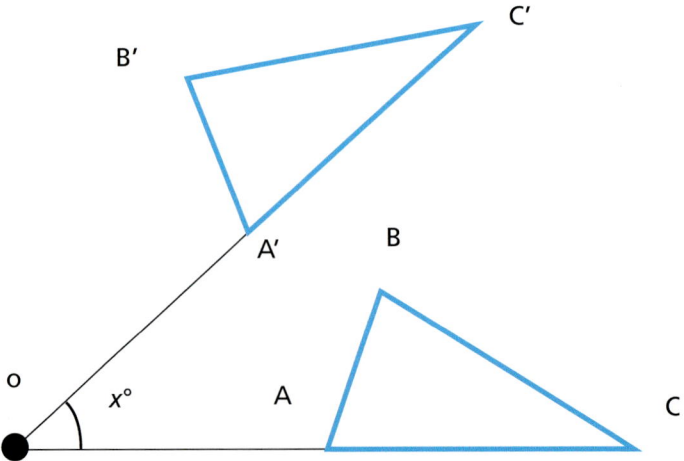

Symmetry

The use of the mirror line gives us a link into the idea of symmetry. The form of symmetry here is called reflective symmetry. The best way to explore line symmetry is through folding shapes. If you can fold the shape in half so that the pieces fit exactly on top of each other you have found a **line of symmetry**. For example,

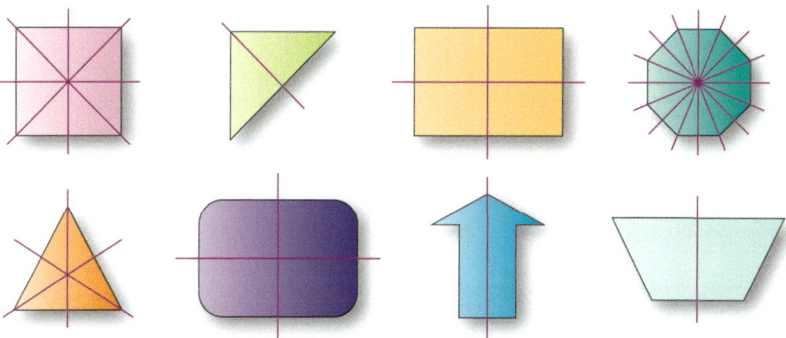

A shape can also be described as having **rotational symmetry**. This means that you can rotate the shape and it will look the same before you have rotated the shape through 360 degrees. The **order of rotational symmetry** is the number of times the shape is repeated before it makes a complete turn, and the **degree of rotational symmetry** is the amount of turn you must make for a repeat to take place.

Portfolio task 8.4

Draw the **regular polygons** given in the table on page 172. Find all the lines of symmetry for each shape. What do you notice about the result? Why do you think this is the case?

Teaching points

Teaching point 1: issues with language – describing properties; positional language

Find a couple of friends, or phone a couple of people up. Ask them to find a scrap piece of paper and draw a triangle. Our prediction would be that they draw an equilateral triangle.

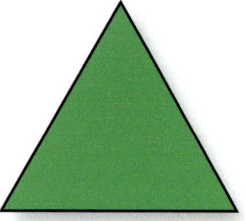

We have tried this out with very large groups of pre-service teachers, teachers and students, and this is often the case. Of course, there is a huge range of triangles to choose from (isosceles, scalene, right-angled), but it appears as though the default option is equilateral. Indeed, many young children find it difficult to see a shape like this as a triangle as it looks very different from the triangles that they have previously seen.

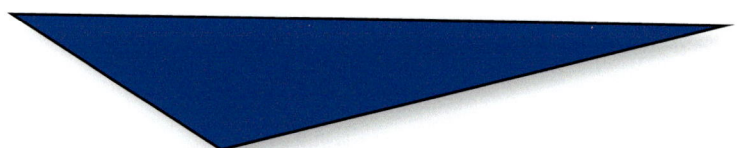

This looks very different from the shape they were first introduced to as "triangle." This takes us back to the discussion about generalisation earlier in the book. When introducing shapes to students it is very important that they see the whole range of shapes. Rather than offering one example of a triangle, offer a whole range and ask students what all the shapes have in common. In this way they will see that what makes a triangle a triangle is having three straight sides all joined together.

Similarly, when introducing vocabulary that may have a range of meanings such as "face" or "edge," take time to find out the students' meanings before imposing a new definition on them. A simple activity to help with this is to ask students to draw or write a definition for a set of key words. These might be *face*, *edge*, *vertex*, *corner*, *line*, *straight*. You can then use the student's definitions to help explain the mathematical definition.

Teaching point 2: orientation of 2D shapes

Tony once carried out some work for an examination board looking at scripts for the post-16 examination in England. One of the criteria that was used to help decide borderline cases was whether or not the students could use Pythagoras' theorem effectively. It was clear from looking at the scripts that if a triangle such as this

appeared in the question, most students realised that they should try to use what they knew about right-angled triangles. However, when faced with a triangle such as this,

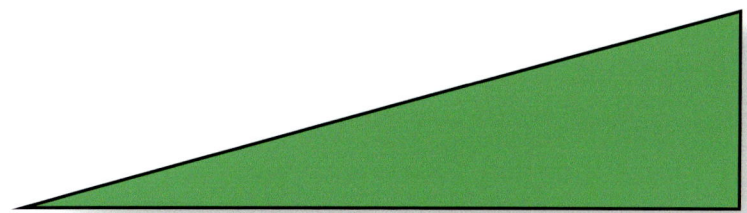

students seemed not to recognise the shape as a right-angled tri-angle. This is a similar issue to the one mentioned in teaching point 1. Students get used to seeing a single representation of a shape and cannot transfer their learning when faced with an unfamiliar version. An activity such as "Shapes in a bag" can help with this.

Get yourself a cloth bag and a set of large shapes. Gradually and slowly reveal a shape. When the students can only see a small portion of the shape, ask what different shapes it could be, slowly reveal a little bit more, and ask again. As you reveal the shape it will become clear what the shape is, but the students' focus is on noticing the properties of shapes rather than simply remembering names.

Teaching point 3: 2D representations of 3D objects

Try to sketch a sphere. Did you draw

rather than something like this?

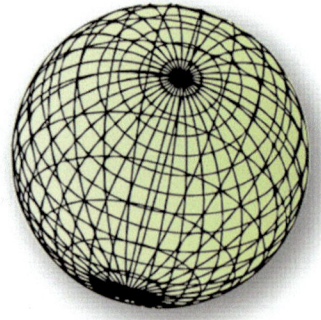

Students find it very difficult to sketch 3D shapes. It is difficult after all – in trying to draw a 3D object using only two dimensions we will often draw representations such as a square for a cube. This can lead to difficulties when working with more complex shapes, and the ability to sketch 3D objects supports us in visualising.

The following activity introduces isometric paper as a way of sketching 3D rectangular prisms. It also raises the importance of orientation. We need to look at 3D shapes such as this in different ways if we are to be able to reproduce them exactly. As the example at the beginning of this chapter suggested, sometimes students who can quickly grasp sketching on isometric paper may not have as quick a grasp of other areas of mathematics.

This is an important skill, so it is good for these students' views of themselves as mathematicians to see themselves as the experts in this particular area. A helpful extension of this activity is to draw the plan view and the two side elevations of each shape. These are the two-dimensional views that you get if you look directly down onto the shape and directly at the front and the side.

Building shapes out of cubes

You will need:

- interlocking cubes
- dotty paper.

1 Build each of these shapes using the numbers of cubes shown
   (a) 8 cubes   (b) 10 cubes   (c) 9 cubes

 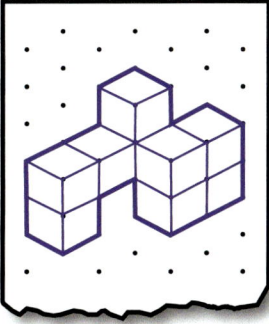 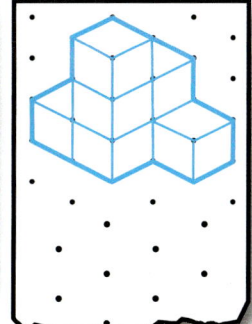

2 Can you find more than one way to build each shape?
   a Make this shape.
   b Add a cube to make a new shape
   c Draw your new shape on dotty paper
   d How many different shapes can you make using the extra cube?

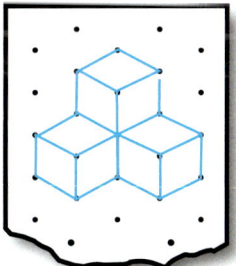

3 a Make this shape
   b Add a cube to make a new shape
   c Draw your new shape on dotty paper
   d How many different shapes can you make using the extra cube?

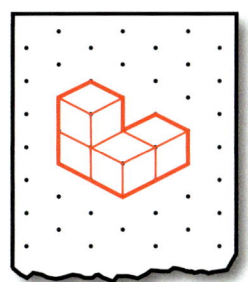

Teaching point 4: errors in not seeing angle as dynamic

An early activity that you will work on with students is to ask them to order angles intuitively in terms of their size. This follows discussion of angle as a measure of turn using strips of card pinned at one end.

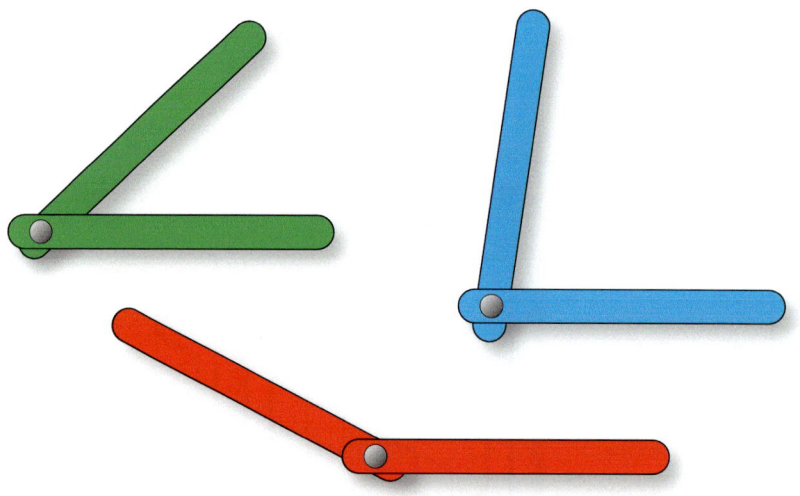

Alfie was looking confused when he was observed exploring this activity. He looked at this angle

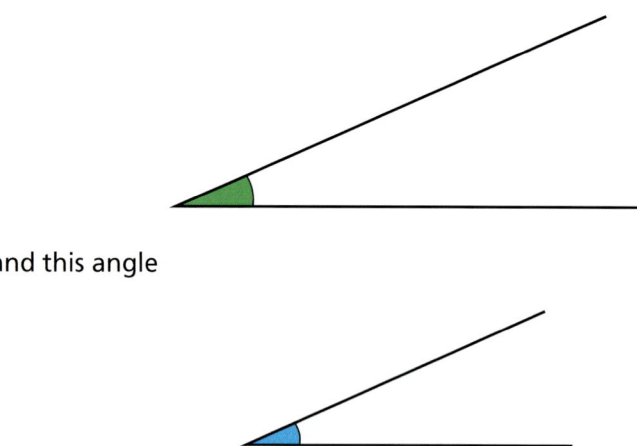

and this angle

and said, "The first angle is much bigger but it's the same size." He was confused as he was noticing the length of the lines and equating that with the size of the angle as well as realising that angle is a measure of turn. He was able to be convinced that they were actually the same angle by extending the **rays** on the second angle so that it appeared identical to the first angle.

Teaching point 5: estimating and measuring angles

There are two common mistakes when using a protractor to measure angles. The first error is using the wrong scale to measure the angle. The two scales are often referred to as the "inner scale" and the "outer scale." Asking students to estimate angles before measuring them will help them decide which scale they need to use.

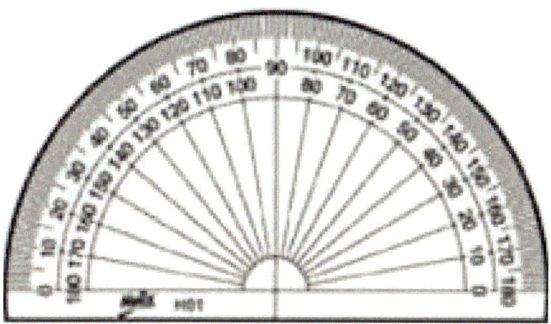

So, if we were measuring an angle of 63° using this protractor, we would know that we should be looking for "about" 60° rather than using the wrong scale giving an answer of 123°.

The second error to look out for is when students place the bottom of the protractor on the vertex rather than the point where the "0–180" line and the 90-line cross.

The first angle that most students are introduced to is a right angle or 90°. Using this fact, we can help students estimate many other angles. They will be able to see that two right angles, which make a straight line, are 180°. Similarly, half a right angle must be 45°. Also, if they divide a right angle into 3 they will get an angle of 30°. It is helpful to ask students to sketch as many different angles as they can, given angles of 30°, 45° and 90°. They can then classify these angles as acute and obtuse. If they get used to having a sense of angle this will help them when they are using protractors to measure angles to 1° of accuracy.

Teaching point 6: the language of coordinates

Some teachers remind students to, "Go along the corridor and then up the stairs" as a way of reminding them to write down the x-coordinate first and then the y-coordinate. Many students remain unconvinced that it really matters in which order you write the coordinates and so seem to forget. Tony once observed a teacher convince children that order matters and that the positive and negative signs are important when working in four quadrants. He drew a set of axes on the board and marked a point A in the first quadrant. He then asked the group, "What shall I call the point A?" One of the group immediately shouted out "Alisha." Rather than show any irritation the teacher asked another student to come up to the front, drew another set of axes and asked them to draw "Alisha." The student said, "But it could go anywhere." The teacher then asked for other ways to describe the point. Students responded with answers such as, "It's 2 along and 3 up," or "Its x-coordinate is 2 and its y-coordinate is 3." This allowed him to draw up all the possible ways of naming the point, illustrating the importance of coordinates as giving information.

Teaching point 7: confusions with mirror lines

There are a number of errors you may notice students making when using mirror lines to explore symmetry. Earlier in the chapter we suggested asking students for their own definitions of key terms as a way of finding out what their current understandings are. A group of Year 3 students were asked what "mirror line" meant to them. Ben suggested that "it is something that splits a shape in half." This was accepted by the group. The teacher than asked the class to draw the lines of symmetry on this rectangle.

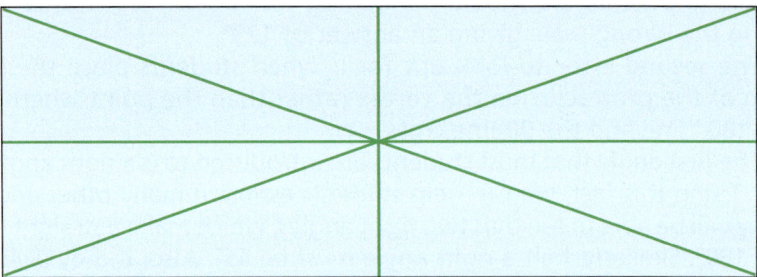

Again, they were all happy that this showed a rectangle had four lines of symmetry. It was only when they were asked to try to fold the rectangle along the diagonals so that the halves "matched up" that they agreed these were not lines of symmetry. (Try it if you don't believe us!) So, the class returned to their definition of a mirror line and said, "It splits it in half so that each half folds exactly onto itself."

Similarly, if a shape is to be reflected in a mirror line and the original shape is placed some distance away from the line, students often place the reflection flush to the line. Simply remind them that the "reflected" shape should be the same distance from the mirror line as the original shape. Whilst we would suggest using practical materials in this case, we would advise against actually using a mirror to explore line symmetry. Mirrors are often very difficult for students to use effectively. Avoid using mirrors at this stage and reinforce the importance of reflections being equidistant from the mirror line. Students benefit from plenty of practice in drawing shapes and their reflections.

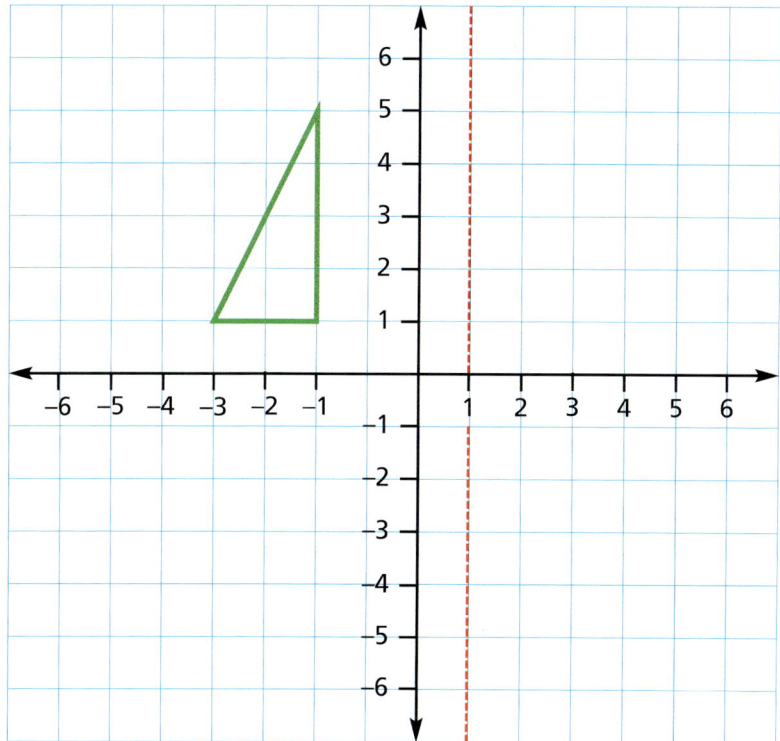

As well as providing students with a starting shape and a mirror line in which to reflect it, it is helpful to give students an initial shape and a reflected shape and remove the mirror line. The task is to find the mirror line. This helps reinforce the notion of equal distance.

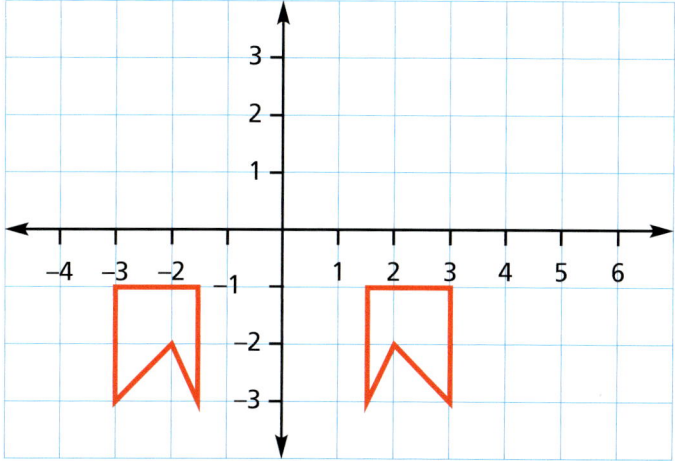

Teaching point 8: rotations about vertices and centres of shapes

As with all areas of geometry it is important to focus on the dynamic. If you are teaching "rotation" to a group of learners, they need to rotate a shape about either the centre of the shape or one of its vertices and then sketch the result. In that way they will be able to describe what is happening for themselves rather than simply trying to remember "rules" they have been told.

A useful strategy is to ask the students to work in pairs or groups so they can check each other's answers. They very quickly see if they are getting the same results, and, with encouragement, support each other in identifying their errors. Encourage students to constantly articulate what they are doing so that others can repeat the process accurately. It also helps students consolidate their own understanding of the process.

In practice

The following lesson plan and evaluation describe a lesson taught to a group of Year 3 students who have some understanding of the language of shape. I wanted them to develop their language in order to be able to identify and describe the properties of shapes in greater detail.

I also wanted them to become increasingly sophisticated in their use of technical vocabulary. The particular focus for this lesson was developing skills in visualising and naming 2D shapes and describing their properties.

The open questions meant that I was able to see what their current understandings were and so I could model more sophisticated vocabulary in my "interpretation" of their answers.

Learning intentions

To classify regular and irregular two-dimensional shapes
To describe regular and irregular two-dimensional shapes and their properties

Key vocabulary

circle, triangle, square, rectangle, hexagon, parallel, perpendicular, corner, curved, straight, sides, angles, same/different

Resources

Collection of shapes

Context for lesson

This is my first lesson with this group of students exploring their understanding of shape. My aim was to discover the students' prior knowledge to support me in developing their understanding of properties of 2D shapes.

Warm-up activity: shapes in the bag

Students sat in pairs to play the "shapes in the bag" game. I gradually revealed a shape and the pairs had to discuss what the shape might be. Each pair had to justify their guess by describing a property of the shape they were guessing. On each occasion we explored all the shapes that it might be as well as the shapes that were not possible. For example, "that can't be a square because it has got curved lines."

Launch

I asked the whole group to sit in a circle with all the shapes in the middle of the circle. I asked one student to sort the shapes in any way that they liked. The others in the group had to guess what the criterion for sorting was. If they guessed correctly, they took over and had to sort the shapes according to a different criterion. This allowed me to point out the different properties of shapes and model the geometrical language that I wanted to encourage the students to use.

Explore

The students were placed into mixed-ability groups of three or four and continued to play the game until each member of the small group had taken a turn to sort the shapes. After each students' turn, the group had to record the criteria used to sort the shapes.

The next activity involved one in the group describing one of the shapes so that the others in the group could sketch it in their books. As I moved around the groups, I made a note of the range of vocabulary being used during both activities.

Summary

The students returned to the floor and I strategically selected a few students to share the different ways that they sorted the shapes. This sharing of ideas was also the starting point for our class vocabulary display.

Rationale and evaluation

My aim was to discover the students' current knowledge, and I was pleased that I left the lesson with a clear understanding of the range of prior experience. Some of the students are secure in their knowledge of triangles, squares and rectangles; Bartek even knew the difference between isosceles and equilateral triangles. However, Malc thought that the scalene couldn't be a triangle because it was "all wonky." I will introduce a group to the different sorts of triangles next time through a classification exercise.

Most of the group found the irregular shapes difficult to work with – they are obviously more used to working with regular shapes. I will start every day this week with an example of an irregular shape and a regular shape so that they get used to seeing irregular shapes.

The students listened carefully to each other, and I could see that some of the children who had a more developed vocabulary were using it, almost to show off, but I was pleased that the other students were exposed to a wide range of vocabulary. I made a rule that if anyone didn't understand a word they had to ask and that the person who had used the word had to explain what it meant by drawing an example.

I made sure to ask students what they meant when they introduced some new vocabulary. The students were very impressed with the long list of vocabulary that we compiled together at the end of the lesson. You could see a sense of pride in their knowledge of the vocabulary of shape.

Assessing geometry

A useful assessment task is to ask your students to classify shapes in a number of ways. Use a wide range of concrete shapes that they can sort physically. These are available from many suppliers of mathematical resources. Using sorting grids is a useful technique as this requires the students to look at properties very carefully. You could use properties of symmetry. For example

	No rotational symmetry	Has rotational symmetry
No lines of symmetry		
One line of symmetry		
Two lines of symmetry		
More than two lines of symmetry		

or ideas of regularity

	Regular	Irregular
Triangle		
Quadrilateral		
Pentagon		
Hexagon		

This allows you to focus on the wide range of varieties of polygons that exist and use specialist language where appropriate. So, a regular triangle is an equilateral, a regular quadrilateral is a square and so on. This same activity can be adapted for three-dimensional shapes.

Cross-curricular teaching of geometry

The idea of this section is to draw on the key ideas within the chapter to develop a cross-curricular project that you can explore with your learners over a series of lessons. This allows you and your class to develop your subject knowledge together.

Activities drawing on design and packaging are ideal for developing into a cross-curricular project for your class. This sort of activity is particularly useful for an end-of-term celebration. Perhaps the school is holding a fete and needs to award prizes or wrapped up surprises for some of the games. The students need to work out how they can design the nets that will "fold up" to make the packages that will hold the prizes as securely as possible. They will also need to visualise the 3D shape that will result from the 2D net so that they can decorate the net before making it up into the package.

Other possibilities are designing patchwork quilt. In her blog, Marie-Therese Wisniowski suggests that Aboriginal women were the first Australian women to make patchwork quilts. Possum and kangaroo pelts were cut into rectangles and sewn together to form cloaks and blankets. Creating a class quilt to given dimensions with each student contributing leads to lots of measuring as well as creativity, especially if you create a crazy patchwork like this one.

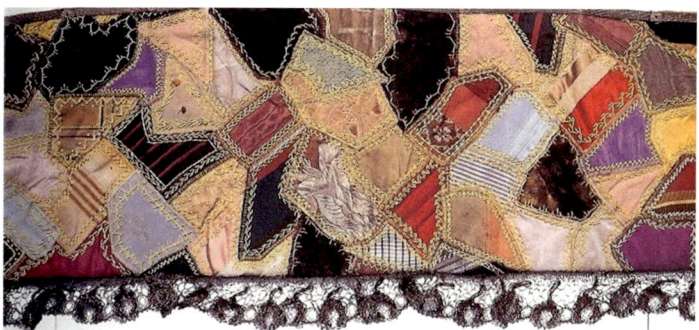

A final possibility, which also asks the students to use the idea of tessellations, is to calculate how much it would cost to make and hang bunting around the classroom. Tessellation is the process of covering a two-dimensional space with repeating two-dimensional shapes. Most individual pieces of bunting are isosceles triangles so the students will need to work out the most efficient way to cut these out of pieces of material. There is also much scope for measurement to differing degrees of accuracy in this activity.

Summary

This chapter has shown you how the ideas of visualisation, categorisation and position underpin geometry. To be able to understand geometrical ideas you need to be able to see and manipulate shapes in your "mind's eye." This is a skill that can be developed. You may need to develop it in yourself before you can work with your students. Or it may be that by helping your students to become better at visualising you become more skilled yourself. The chapter opened with an example of a visualisation activity you can try with your students. It is well worth building your own bank of these activities and using them regularly.

There has been a focus on the properties of shapes. A knowledge of the range of possible properties is important in order to be able to classify shapes. So, another area to concentrate on with your students is developing their vocabulary of shape. Use class-created anchor charts to make sure the vocabulary is clearly displayed so it becomes a part of the language of the classroom. It is important that you model the language of shape as often as possible, in the same way that you will often use number facts when taking a register; you can use the language of shape whenever you notice new shapes, whatever the subject.

Reflections on this chapter

Do you see the learning of geometry in a slightly different way than you see learning and teaching aspects of number? Sometimes students do not seem to regard the topic as "important." But it is, not only in terms of being successful in assessments which all include shape but also as a way of viewing and describing the world. You may well have discovered a new skill, that of visualisation. If you cannot visualise effectively, maybe as a result of not learning it at school, this will inhibit you when you are trying to solve practical problems around the home, such as deciding how to pack the car boot or arrange furniture a room in the best way.

Learning about geometry also seems to motivate children. They enjoy talking about the shapes, they enjoy making new shapes, they enjoy describing what they are seeing. We hope that you feel able to draw on all the practical aspects of shape in your work with your students. Do not forget that geometry is a dynamic and practical area of mathematics, and we cannot learn about it without being dynamic and practical.

We also hope that you feel the language of shape has been demystified. It isn't complicated, but there is a lot of vocabulary to learn and remember. However, this is just a matter of regular use. Talk about shape with your students on a daily basis. Have a shape of the week and add a property a day. Play the "shapes in a bag" game whenever you have to fill five minutes at the end of a lesson, and soon you will all have really well-developed vocabulary of shape.

Going further

Mindstorms

SEYMOUR PAPERT

This seminal text brings together Papert's vision for the ways in which computers and ICT can transform the ways that children learn and how the programming language of Logo, which is predominantly used to explore shape and space, can be utilised to support children in developing independent thinking skills. Whilst this is a research text, there are lots of immediately applicable ideas.

Papert, S. (1993) *Mindstorms: Children, computers and powerful ideas.* London: Basic Books.

Answers (to question on page 167)

Possible properties to sort the shapes by. Of course, there are many more possibilities. These are not intended to give the "correct" answer:

- opposite sides parallel
- opposite sides congruent
- at least one obtuse angle
- at least one right angle
- all sides congruent
- all angles congruent
- two consecutive sides congruent.

Solution to the "name the polygons" task:

- parallelogram
- quadrilateral
- regular polygon
- opposite angles congruent
- pentagon
- hexagon
- octagon
- rhombus
- isosceles
- trapezoid
- concave polygon
- convex polygon.

Self-audit

Carry out this audit to explore how your learning has progressed as a result of working on the ideas in this chapter. Include the results in your portfolio.

1 Draw four sets of axes in all four quadrants with the coordinates labelled up from −5 to +5 (see page 171). For each question you should draw the initial shape on the set of axes in one colour and the second shape in another colour. Then write down instructions which explain how to move from the first shape to the second shape. You should use at least two sets of instructions each time and include reflections, rotations and translations if you can

   a   initial shape, right-angled triangle (− 1,1), (− 1,5), (− 5, − 1); dotted shape (1, − 1), (5, − 1), (1, − 5)

   b   initial shape, rectangle (− 2,3), (− 2,5), (− 5,5), (− 5,3); dotted shape (0, − 2), (0, − 5), (− 2, − 5), (− 2, − 2)

   c   initial shape and "L" shape (2,1), (5,1), (5,4), (4,4), (4,2), (2,2); dotted shape, rotate through 90° about (2,1) then translate 6 in the direction −X

Now draw your own starting and finishing shape together with instructions of how to get from one shape to the other.

2 Draw four shapes. They should include at least one triangle, one quadrilateral and one pentagon. Underneath the shape write the instructions to help a friend draw the shapes. You will need to include the length of the sides and the approximate angle at each vertex. Underneath the shapes write down a list of each shape's properties and the shape's name.

CHAPTER 9
MEASUREMENT

Introduction

All of the activities you have been introduced to in this book are versions of activities we have used in our own teaching. We think that you learn most about how to teach mathematics from teaching mathematics. And that there is always more to learn. For example, Tony was working with a group of six- and seven-year-olds using balance scales and exploring the relative **mass** of beans and pulses when he noticed one of the teaching assistants using her hand to measure out pulses. He talked to her later, and she explained that when she was cooking she used her hand, cupped to a greater or lesser amount, as a measure that would compare to a teaspoon or a tablespoon or any measure in between. He realised that this was a "unit" of measurement that he had not seen used before but that many of the children he worked with would be used to. This reminds us of the importance of learning from the cultural diversity within the classrooms that we work in.

As with activities involving shape and space, activities involving measuring allow us to notice students excelling, who have not always succeeded immediately when working with number. You might notice those students who have a well-developed sense of spatial awareness coming to the fore and that those who have experience outside the classroom of constructing or cooking are able to bring this prior experience to bear. Those students who sometimes struggle to remember their multiplication facts can easily estimate how long a shelf needs to be to fit into a space or can quickly stack building blocks into a box, another reminder that we should not use the term "ability" in an unthinking way.

DOI: 10.4324/9781003315155-9

Reflecting on the story with which we opened this chapter, it is worthwhile pausing for a moment to think about the measures we have a good sense of. Jess cycles as her primary mode of transport, so she has an appreciation for those sorts of distances. It is 12km from her house to work, 3km to the closest shops, and so on. In terms of volume, depending on what state you are reading this in you may be more familiar with the size of a pot or a middy, a schooner or a pint. The starting point to this chapter reflects on which units of measurement we have a good understanding of and those that we cannot immediately picture.

Starting point

Portfolio task 9.1

Write a list of the units of measurement you have a sense of, as we did earlier. Try to complete the following list. What do you think weighs about 1kg, has a capacity of about a litre or measures about 1cm?

1mm	1g	1ml
1cm	1kg	10ml
1m	100kg	500ml
1km		1 litre
10km		100 litres

Which measurements did you find easy and which did you find hard?

Talk to your friends about it and think about how your prior experiences have informed your knowledge of measurements.

Measuring is not just about units of measurement; other key skills are reading scales and measuring instruments, including clocks, and understanding concepts such as area and perimeter. We will explore all of these areas in detail throughout the chapter. First, let us see how children's understanding of measuring is developed during their time in primary school.

TAKING IT FURTHER: **FROM THE RESEARCH**

Derek Hurrell from the University of Notre Dame Australia explored how learning and teaching measurement can be used to support students in developing more general mathematical skills and aptitudes. In the paper *Measurement: Five Considerations to Add Even More Impact to Your Program*. Available at https://researchonline.nd.edu.au/cgi/viewcontent.cgi?article=1171&context=edu_article he shows how the mathematics of measurement can be connected to other areas of mathematics.

The key ideas he explores and that he argues are key ideas across mathematics and not just for measurement are:

Identifying what we want to measure as there are so many attributes of an object or space that can be measured.

Comparing and ordering satisfies our initial, natural inclinations and then creates a desire for precision.

Using non-standard units allows the learner to explore convenience and appropriateness of units relative to the attribute being measured.

Using standard units is necessary to be able to perform calculations and to communicate measurements to others.

Applying formulae of measurement is a product of the processes and aforementioned understandings, not a replacement.

He concludes by proposing that every mathematics lesson is a measurement lesson as it is rare that we will engage with mathematics without estimating.

The next section outlines the development of children's understanding about measurement as they move between foundation stage and the end of their time in primary education. You should keep in mind the advice from the previous research that as much of this as possible should be explored actively and dynamically. As Derek Hurrell says, "A measurement lesson should not be a passive lesson."

Progression in measuring

Foundations for measuring

Initially students will begin to understand the language of measurement and comparison using vocabulary such as "greater," "smaller," "heavier" and "lighter" to compare quantities. They will compare relative lengths, mass and capacity. To introduce children to ideas of time, you should also use vocabulary such as "before" and "after." So, at this stage children should be engaging in a wide variety of practical activities that allow you to develop this vocabulary with them; this will include linking the names of days of the week with events that take place on those days.

Beginning to understand measuring

At first students should carry out activities that involve measuring and weighing in order to compare objects. They will use suitable **uniform non-standard units** and then **standard units** for this comparison. (In non-standard units, uniform means that we use something that has a uniform measurement, so we can measure length using multilink cubes or use a fixed number of wooden blocks to compare weights. Standard units are those in common usage, such as metres, litres, kilograms and all the related units.) Once children have an understanding of using non-standard units to compare, they can start to use a range of measuring instruments such as metre sticks and measuring jugs. Students will continue to estimate, compare and measure but now relying on standard units. They will be able to choose suitable measuring instruments to help them. At this stage you should introduce them to scales and teach them how to read the numbered divisions on a scale, such as a weighing scale or measuring jug. They will also be using a ruler, which is another scale, to draw and measure lines to the nearest centimetre.

To build students' sense of time you can introduce them to vocabulary related to time such as days of the week and months. They can also begin to tell the time to the hour and to the half hour. You will develop their sense of time using seconds, minutes and hours. They will then be able to tell the time to the nearest quarter of an hour and be able to identify time intervals even across an hour. So, for example, they will know that there are 15 minutes between 5 to 6 and 10 past 6. Once they have a good grasp of telling the time to 15 minutes, you can work on telling the time to the nearest 5 minutes. They should use both analogue and digital clocks.

Becoming confident in measuring

At this stage students will come to understand the relationships between the standard units, kilometres and metres, kilograms and grams and so on. They will be able to record their measurements using appropriate units. They will develop their skills reading scales, becoming able to read them to the nearest division and half division, including reading scales that are only partially numbered. As a next step, they will learn the abbreviations for standard units and will understand the meanings of "kilo," "centi" and "milli." The basic units are metre, litre and gram. These are the basic units as they do not have a prefix. The prefixes tell you what multiple of the unit you require. So "kilo" tells you that you need 1,000 of the unit; "centi" tells you this is 1/100 of the basic unit and "milli" is 1/1000. This leads on to decimal notation, so your learners will understand that 1.5 metres is the same as 1m 50cm. They will be reading scales and recording readings to the nearest tenth of a unit.

Students will be reading the time on 24-hour digital clocks and to the nearest minute on an analogue clock, as well as measuring time intervals so they can work out beginning and end times for given time intervals. For example, they will know that if a TV programme lasts 30 minutes and starts at 9.40 it will end at 10.10. Once they have an understanding of this, students will move on to be able to read time-tables and time using the 24-hour clock and use a calendar to work out time intervals.

You will also introduce measurements linked to rectangles, including measuring the **perimeter** and finding the **area** of **rectilinear shapes** by counting squares in a grid. The perimeter of a shape is the distance all the way around its outside edge. This is measured in mm, cm or km. The area of a shape is the amount of space it takes up. This is measured in square mm, square cm or square km, written mm^2, cm^2 or km^2. A rectilinear shape is one that can be split up into a series of rectangles.

Developing these ideas, students will be able to convert larger to smaller units using one place of decimals. For example, they will be able to change 7.3kg to 7,300g. They will be able to interpret readings that lie between two divisions on a scale and draw and measure lines to the nearest millimetre.

They will also be calculating the perimeter of regular and irregular polygons and using the formula for the area of a rectangle to calculate area. A **polygon** is any shape with straight edges. A **regular** polygon must have all sides the same length and all the angles between the sides the same, otherwise the polygon is **irregular**. The formula to find the area of a rectangle is length multiplied by width.

Extending understanding of measuring

At this stage students will be able to read a wide variety of scales and will understand that the measurements made on scales are only approximate to a given **degree of accuracy**. They will be able to calculate perimeters and areas of rectilinear shapes.

Developing these skills students will operate up to three decimal places knowing that 2,541ml is the same as 2.541 litres. They will apply their knowledge of measuring to solve a range of problems and will also be able to calculate using **imperial units** (such as pints and ounces) still in everyday use, having an understanding of their approximate metric values.

This section illustrates how, from the building blocks of the foundation stage where children are developing the language of comparison, they progress to the stage where they are accurately estimating, measuring and comparing; reading any scale accurately; telling the time and calculating area and perimeter of rectilinear shapes. The next section outlines the big ideas and key skills that underpin measuring.

Big ideas

Conservation and comparison

We measure in order to compare objects. We need to see if something is the right length to fit in a particular space or if it will hold objects of a certain length. One of the earliest principles students come to understand is that the measurements of a particular object stay the same wherever we place it. Young children may describe something as longer than something else if they are not aligned.

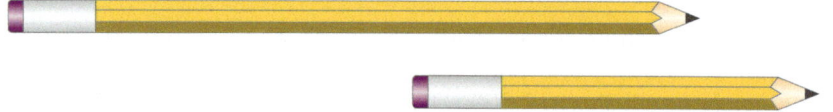

In this case they may say that the shorter pencil is "bigger" or "longer" than the other. Young children need to use non-standard units such as pencils, wooden rods and cubes to come to an understanding that measurement is **conserved** (stays the same) before they move on to using standard units. These difficulties with ideas of conservation are discussed in more detail later.

Units of measure

Units of measurement have been agreed by mathematicians and scientists in order to make measurement consistent throughout the world. In Australia metric units, first introduced in 18th-century France, are the system taught in schools. Metric units are in common usage as they are based on powers of ten and so are easier to work with. Students may also come across imperial units that are used in the UK, the US and most of the Caribbean. The most common imperial units are:

Length	12 inches in 1 foot; 3 feet in 1 yard; 1,760 yards in 1 mile.
Area	144 square inches in 1 square foot; 9 square feet in 1 square yard; 640 acres in 1 square mile (an acre is about 4,840 square yards).
Volume	1,728 cubic inches in 1 cubic foot; 27 cubic feet in 1 cubic yard.
Capacity	5 fluid ounces in 1 gill; 4 gills in 1 pint; 8 pints in 1 gallon.
Mass	16 ounces in 1 pound, 14 pounds in 1 stone; 8 stones in 1 hundredweight (cwt); 20cwt in 1 ton.

Looking at this list you can probably see why the decision was made to focus on metric units! Metric units are sometimes referred to as SI units (Système Internationale d'Unités). The table below shows the SI units together with the conversions to imperial units.

Attribute	SI unit	Abbreviation	Imperial units	Abbreviation	Conversion
Length	Metre	m	Inches, feet, yards, miles	in, ft, yd	1in = 2.54cm 1ft = 0.3m
Mass	Kilogram	kg	Ounces, pounds, stones	oz, lb, st	1oz = 28.35g 1lb = 0.45kg
Time	Second	s			
Area	Square metre	m²	Square inches, acres	sq in	
Volume	Cubic metre	m³	Cubic inches, cubic feet	cu in, cu ft	
Capacity	Litre	l	Pints, gallons	pt, gal	1pt = 0.56l

As you read in the opening section to the chapter, metric units are easy to use as the prefixes (kilo, milli, centi) tell you the conversion between the units.

Scales

All measurement is against a scale based on the unit we are using for comparison. Measurement is against a continuous scale, that is, we are always measuring approximately. If we say a line is 3.2cm long, what we are actually saying is that it is 3.2cm long to the nearest tenth of a centimetre. It may be 3.21cm or 3.19cm, but we have decided that saying 3.2cm is accurate enough. Reading scales is one of the key skills to teach children. It is also useful to teach them how to create their own scales to measure. This helps them see the importance of standardisation and also the approximate nature of measurement.

Teaching points

Teaching point 1: conservation of mass and capacity

In the "Big ideas" section earlier in the chapter we suggested that young children do not always remember that you need to place the ends of two objects together to compare length. Similarly, if you move an object so that it is in a different orientation, the students may need some convincing that the length of the pencil has remained the same (or been "conserved").

Before we explore students' understanding of conservation of mass it is worth defining mass. In everyday usage, mass is more commonly referred to as weight, although the precise scientific definition is the strength of the gravitational pull on the object, that is, how heavy it is. The distinction between mass and weight is important for extremely precise measurements that may be affected by slight differences in the strength of the earth's gravitational field at different places and for places far from the surface of the earth, such as in space or on other planets. This means that you can convert exactly between weight and mass on the earth's surface. This confusion between mass and weight is heightened by the fact that in much of the metric world, weight is not dealt with, and mass is used in its place almost exclusively. The main difference is that if you were to leave the Earth and go to the Moon, your weight would change but your mass would remain constant.

Conservation of mass is an even more difficult concept for young children. An activity you can use that begins to get children thinking about this is to put four or five boxes of different shapes and sizes on a table and ask the group to place them in order of mass. Deliberately place the heaviest object in the smallest box. Once they have ordered the boxes, get the group to compare the masses of the boxes by lifting them up. They will be amazed that the smallest box is the heaviest as they have the misconception that the mass of an object is directly linked to its volume.

Another activity that deals directly with this misconception is to take a large spherical piece of play dough and place it in a balance so that it exactly balances with a counterweight. Then move the play dough from the scale and roll it flat. Ask the group if they think it will be lighter, heavier or the same. Students will often think that it will now be lighter, as it is thinner. They will be surprised that this is not the case.

Another example of this is to place three or more containers in front of the students. The containers should range from very wide-based containers to long thin ones. Pour the same amount of liquid into each container. The liquid will not come very high up the wide-based container, and the same volume may fill the container with the smaller base.

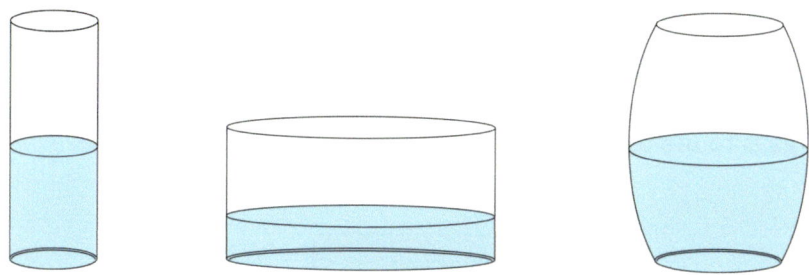

Lots of practical experience of weighing a wide range of objects and measuring capacity are the most effective way to deal with this misconception.

Teaching point 2: conservation of area

A small group of children were finding the areas of rectangles by counting squares. After they had agreed that the area of this shape was 18cm².

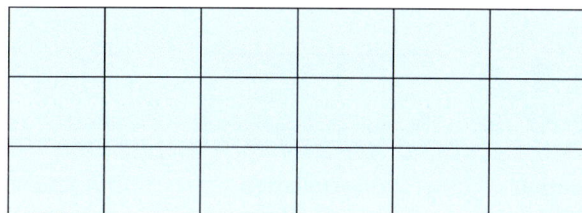

The group was asked to make new shapes by cutting across diagonals of the squares. They created three new shapes, all including triangles:

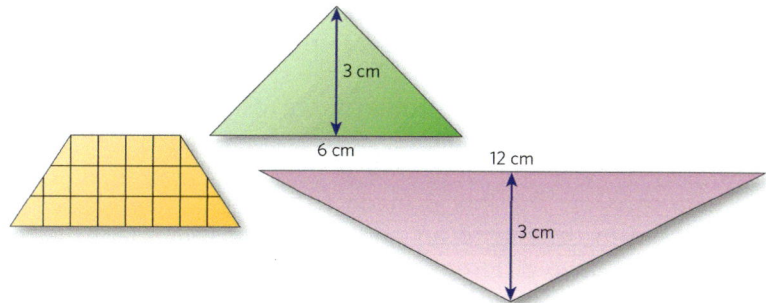

When they were asked what they thought the areas of these shapes would be, none of them believed that the area would still be 18cm². In fact, one student had to count each one three times to convince herself. She said, "That can't be right. Look, this one is much more spread out and so it should have a bigger area." In her view, the fact that one of the shapes was longer suggested that it should have a greater area.

A good way to develop children's understanding of conservation of area is to use tangrams

A tangram

or other tessellating activities

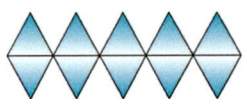

A tessellation of triangles

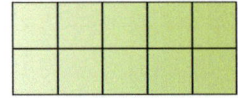

A tessellation of squares

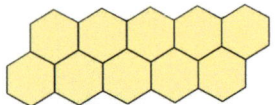

A tessellation of hexagons

to create a wide range of shapes with the same area. A tangram is a Chinese puzzle consisting of a square that is cut up into seven pieces. These pieces can be used to make many different shapes. As these are all made from the same pieces, they will all have the same area.

In the following activity students are asked to rearrange 4cm squares to make different shapes. Although the shapes look very different, they all have the same area as they are all made up from the same 4cm squares. The link is made to perimeter as a way of supporting them in coming to an understanding that shapes with the same area can have different perimeters. This also consolidates the student's understanding of perimeter.

AREA AND PERIMETER

If two 1cm squares are put together, there is only one way to link them:

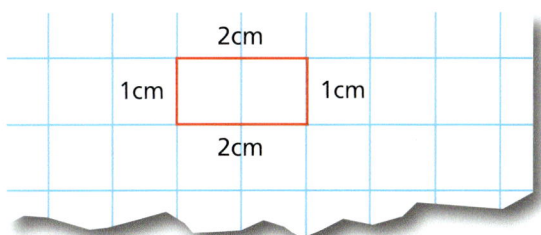

The perimeter is (2 × 2cm) + (2 × 1cm) = 6cm

If you put four squares together, you can make these shapes:

A

B

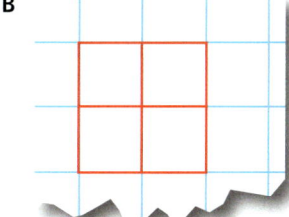

C

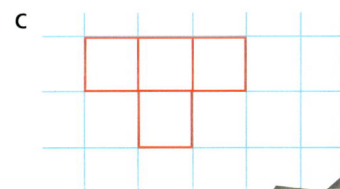

D

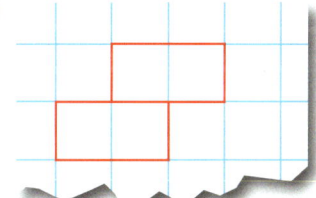

1   Find the perimeter of each of these shapes
2   There is one more four-square shape you can make. Make this shape and record its perimeter
3   Which four-square shape has the shortest perimeter?
4   Use six squares. How many different perimeters can you find?
5   Repeat with 8 squares, then try 10 and 12 squares. Find all the possible perimeters for each number of squares
6   Look for rules or patterns to predict which shape gives the longest perimeter and which shape gives the shortest.

Portfolio task 9.2

A similar activity for more experienced learners can be set as the following open-ended investigation.

Explorers in a gold field are each given 100m of rope and 4 stakes. Each explorer uses the 4 stakes and the rope to section off the piece of land they will dig in order to find gold.

What are the dimensions of the shape they should make to enclose the largest area they can?

Teaching point 3: using appropriate units

The activity that opened the chapter was designed to start you thinking about appropriate units. It can be very confusing that length can be measured in millimetres, centimetres, metres and kilometres. Then, when we measure area, we can use the equivalent measures, although we tend mainly to use square centimetres and square metres. The best way to support students getting used to appropriate units is to use contexts with which they are familiar and to ask for appropriate units rather than asking them to carry out the measuring activity. In this way they can focus on the units, rather than the act of measuring.

Teaching point 4: relationships between dimensions and length, area and volume

Your students may simply try to remember a process when you ask them to solve problems having to do with perimeter, area and volume rather than think through what the problem is asking them to do. You may have heard students say something like, "Is perimeter the one where you add everything together?" or simply multiplying any numbers that they can see on a diagram when asked to find an area, as they remember that you "multiply to find area." Try this activity with a class you are teaching as a way of supporting them in coming to a deeper understanding of the relationship among perimeter, area and volume. In this way, the language of area, perimeter and volume will make more sense to them as well as the relationships among these three ways of describing measures.

Portfolio task 9.3

Decide which of these statements are true and which are false:

- if you double the dimensions of a rectangle you double its perimeter
- if you double the dimensions of a rectangle you double its area
- if you double the lengths of the sides of a rectangle, but leave the width the same, you double the area
- if you double the dimensions of a cuboid you double its volume.

For the false statements write down an equivalent statement which is true. Try to describe why this is true.

Teaching point 5: misreading scales

Students can make a range of errors reading scales. First, you will have noticed students not always measuring accurately because they do not line up their ruler correctly. They may start from the edge of the ruler rather than the zero point (this is not always at the end of a ruler), or sometimes you may notice students starting at the 1cm point and so will add 1cm to the true measurement.

An activity that can help students correct this for themselves is to have a series of lines down one side of a page and a series of measurements down the other side. They have to measure the lines and match them to the appropriate measurement. Avoid using the answer that an error of measuring starting at the 1cm mark would give the student. Any student who is habitually making this error will then have to self-correct in order to match the correct answer.

Students also sometimes forget to make sure that a graduated cylinder or measuring jug is placed on a flat surface when they are reading the scale. They move the cylinder to their eyes rather than their eyes to the cylinder. You need to check that they are placing their eyes level with the liquid so that they read the scale accurately. You can illustrate this to the students by tipping a cylinder and showing them how inaccurate the reading can be.

Teaching point 6: misreading clocks

The best way to support students in learning to tell the time is to have both analogue and digital clocks in the classroom and to refer to them as often as possible. Some students rarely see analogue clocks; they will have a digital reading on their mobile phone or on a TV at home, so you may need to model using an analogue clock frequently.

In *Mathematics Teaching* 209, Rona Catterall describes the way she introduced telling time to her class of six- and seven-year-olds. This is a detailed article about the methods Rona used after she became frustrated with her children struggling with the skills of telling the time.

One example she uses is how you can use a child's understanding of their age to help with telling the time. She reminds us that a child sees their age to be of great importance. So, she asks one of her class how old they are, "I'm six,", they reply. The teacher asks if they can say they are seven? "No," says the student, "I'm not seven until my next birthday." "It's just the same with this clock," says the teacher, with the time set at half past six. "We can't say its seven o'clock until the hand has gone all the way round."

Tony remembers introducing a class to an analogue clock a few years ago and they laughed almost as if he had brought in an ancient timepiece. He set the clock at half past 12 (12.30) as this was lunchtime and asked the group what time was shown by the clock. The range of answers included:

6 past 12

6 to 12

12 and a half

All of these were sensible ways of interpreting a new scale. When we are introduced to a new scale, out of context, we decode it the best we can. In the lesson described previously, Tony decided the best thing he could do was to discover all the possible errors with the students, and so he set them the following task.

Portfolio task 9.4

This clock is showing 3.45.

Write down all the ways you could say this time.

These are some of the answers that the class gave. Explain the errors the children are making:

- 45 minutes to 4
- 9 minutes to 4
- 9 minutes past 3
- 18 minutes past 9.

One of the changes you may noticed over the last few years is that the number of different ways of describing the time has reduced. As people become more used to reading digital clocks and move away from analogue clocks, the descriptions of time are always accurate to the nearest minute. We guess that if you ask students in your class what the time is they will tell you, "It's eleven forty-seven" rather than "just after quarter to twelve." This is an interesting illustration of how technology imposes itself on our day-to-day mathematical practices.

In practice

The lesson plan that follows was used to support a group of children in their understanding of measurement in a Year 3 and Year 4 multi-age classroom. Following the plan is an evaluation of the lesson that explores how successful the plan was in supporting the children to develop their knowledge, skills and understanding.

Learning intentions

To develop estimation skills of length, perimeter, mass and capacity
    To choose and use appropriate instruments and units to estimate, measure and record the mass of various everyday objects

Key vocabulary

hefting, gram, kilogram, volume, capacity, mass, unit of measurement, metric system, matter, density, weight, heavier, lighter

Resources

Scales: kitchen (digital and non-digital), bathroom (digital and non-digital), balance

A range of weights: 10g–1 kg
A range of beakers: 50ml–1litre
Instruments to measure length: rulers, measuring tapes, metre rulers, trundle wheels
Collection of items to measure

Context for lesson

This lesson was carried out towards the beginning of a unit that focused on measurement, specifically mass, volume and capacity. The purpose was to provide a practical experience in estimating, measuring using appropriate scales and accurately recording the weight of everyday objects. It was noted prior to conducting this lesson that students were generally capable of making accurate comparisons through hefting (reporting which object is heavier or lighter) and using a balance scale but lacked experience in using different types of scales. They struggled to interpret the graduations between whole numbers on non-digital scales, and although they could read digital scales, they did not have a sense of the difference between 54.7 g and 54.7 kg, for example.

Warm up activity: beat the J (insert your own initial)

In this activity developed by Rob Vingerhoets from his book *Maths on the Go*, the students hone their estimation skills of a range of measures (mass, length, time, capacity, volume) of various everyday objects. I play this game once a week, mostly as a warmup activity, but it can also be stretched out as a whole session or following the main activity as more of a reflection tool. As the name of the activity suggests, the students' goal is to have closer estimations to the actual measure of the objects presented than the teacher because, let's face it, a competition against the teacher is always very exciting.

I began by choosing two students as the Measuring Monitors. They set themselves up at the Measuring Station and were responsible for the actual measuring of the objects. These students rotate each week so that everyone has the opportunity to be in these roles.

Each student has a booklet of "Beat the J" sheets so that they can track their progress week-to-week, but they could just as easily enter their findings into a simple hand-drawn table in their maths book. In this particular lesson, the items listed under "Object" were: length of a pencil, the mass of an apple, the perimeter of one of the window frames, the capacity of a washing detergent container and the height of the classroom door frame.

These objects were held up, passed around or pointed out, one by one. Students (and myself!) were given time to record estimations next to the item name on our own sheets. Measuring Monitors then chose the appropriate instrument and unit to measure and recorded the actual measurement.

On the board, I wrote "Object," "Actual Measurement," and "Margin of Error" in a table. The margin of error is an important aspect of this activity, as we're trying to improve estimation skills, not achieve exact measurements. For example, the pencil was 14cm long, and so I set the margin of error as +/–3 cm. As the students become more familiar with estimating, feel free to tighten the margin of error. I then held up each item again. This time, Measuring Monitors revealed the actual measurement of each object and students could then work out whether they record a P or an O next to each item according to the margin of error. They tallied up the total number of P and this became their score for this round. Students then compared their score with my score and decided whether or not they beat the J (which some of them did!).

Launch: scavenger hunt

I displayed a large image of a produce scale, one that they would have seen in the supermarket to make the connection to the real world. The students compared it with the scales we had in our classroom, noticing the similarities and the differences, and they were encouraged to ask questions and make general observations.

I then brought out the apple from the warmup and asked several students to come up and weigh the apple using all of the available scales (balance, digital kitchen, non-digital kitchen, bathroom). I guided a discussion around the different benefits and perceived challenges each scale presented when weighing the apple.

Explore: scavenger hunt

I placed various instruments and scales around the room, some relevant to this activity and others less so. We want the students to be able to choose the appropriate scales as part of their discovery and have the opportunity to experiment with different apparatuses. The students recorded their findings individually but were encouraged to work together to find various objects that fit under each heading. The headings were as follows:

An object lighter than 10g

An object between 10 and 100g

An object between 100 and 500g

An object between 500g and 1kg

An object heavier than 1kg

Before they measured the mass of each item they chose for each category, they recorded an estimation.

I knew that this task would not be challenging enough for all students, so I inserted two extra columns on the recording sheets as optional extras: "Actual measurement expressed using a different unit" and "Difference between estimation and actual."

In the middle of the activity, I stopped to address some misconceptions and misunderstandings I was observing, such as leaving out the unit when recording the measurement. This led to a great discussion about the difference between 1.25g and 1.25kg.

Summary: scavenger hunt

Now that the students would have had experience with non-igital scales, I knew I wanted the summary to address interpreting the major divisions and graduations between whole numbers. While the students were completing the scavenger hunt, I took some photos of the scales and used these to display during the summary to prompt discussion. This conversation linked their understandings of fractional thinking and division with measurement. There were four graduations on the scale between 0kg and 1kg, and Levi knew that half of 1 was half and that half of a half was a quarter, but he didn't know how to express it in decimal notation. I then displayed a photo from a digital scale that displayed 354.6g. I asked the group what that would look like on the non-digital scale. The students began making connections and understanding that there are different ways to express the same measurement.

Rationale and evaluation

Because this was a physical, hands-on activity, students were naturally engaged. The students needed that time to explore and play with the mathematics, the scales and the objects. They were curious by the different scales, some of which they had never used before. I roved the room, though mainly I hovered around the non-digital scales, offering assistance in reading the markings between whole numbers. This observation helped informed the summary of the lesson.

In discussing how students found objects for the categories during the summary, Samuel said that he drew on his previous experience playing Beat the J, specifically using the apple as a benchmark when hefting other objects. Amelia talked about how she often helps with the food shopping and had "a good feel" for 1kg because her family often buys 1kg bags of onions or potatoes. That led to a great conversation about how our experiences with measurement will often help us make comparisons and connections.

You will find that playing Beat the __ generates a lot of discussion and can easily lead into many different avenues of learning. In this session, the Measuring Monitors weren't sure how to measure the capacity of the detergent container. There were suggestions from the class, and ultimately the "explicit teaching" came from the students themselves who proposed filling up the graduated cylinders with water and pouring them into the detergent container. Some of the most powerful teaching and learning happens when we hand control over to the students, allowing them to explore and become the experts in the room.

Portfolio task 9.4

Devise a lesson plan which is appropriate for a group of learners you are working with.

The focus should be an aspect of "measuring" that is appropriate to the age group you are working with.

Make sure you think carefully about the context and evaluate the lesson.

Add this lesson plan and evaluation to your subject knowledge portfolio.

Assessing understanding of measurement

The most effective way of assessing your students' measurement skills is to observe them engaging in activities that involve measuring. For early learners this will involve you asking questions about how many small containers your students might think will pour into larger ones, how many hand spans will measure a table or how many strides there will be across a classroom. For older students this can take the form of using containers with a given capacity and estimating the capacity of larger containers using this knowledge. As students become more skilled at measuring using scales you can ask them to create a measuring jug, including calibrating a scale on the jug. You might perhaps ask them to create a 1 litre measuring jug with 50 millilitre divisions given a 100 millilitre measure.

You can assess students' early understanding of time by asking them to order events during a typical day and assign a time to them to the nearest hour. This can become increasingly sophisticated in terms of the accuracy of the time. If you draw on the students' own experience, this is even better. Later you can use timetables to ask students to plan appropriate journeys given set criteria.

Cross-curricular teaching of measurement

As suggested earlier, a useful way both to develop and reinforce the skills of measurement through a cross-curricular project is by asking children to make containers of appropriate sizes. The following two ideas should motivate both your younger learners and those in the later primary years.

Younger learners

Bring three teddy bears into the class. One should be very small, one "medium" sized and one larger. If you want, you could link this to the "Three bears" story. Each group needs to make different objects for the three animals. For example, you might ask one group to make a chair, one a plate and bowl and even one an appropriate shelter. This will involve them in comparing measurements and using the language of comparison.

Older learners

Bring in a range of cereal boxes. The students should find ways of measuring the capacity of the box, and their task is to design a box that is a different shape but that will hold the same amount. Ask them to design the net for the box, including logos, as this means they have to think through how the net fits together to form the final box. An alternative to this is to ask the groups to design and create carrier bags for particular objects.

Summary

The emphasis within this chapter is on the practical nature of measuring and the importance of drawing on students' intuitive understandings and knowledge of measurement in order to develop their skills. The progression section illustrates the importance of developing a language of "measurement" in young children and engaging them actively in carrying out measurement as the only way to learn about "measurement." The key ideas that the chapter then builds on are ideas of conservation and comparison, everyday skills we will use all the time but may not have put a name to before. The teaching points emphasise the fact that we can and need to teach these skills to our students – they aren't simply common sense. Other big ideas such as the units we use to measure and compare objects and the scales we use to support measurement are also described.

Reflections on this chapter

The important thing to take away from this chapter is the small number of skills we are trying to teach our students in order to help them become effective in measurement. The basic foci are "units" and "scales." Students need to have a sense of what a measurement means. If your students leave primary school with an understanding of what a litre looks like, how far 500m is, what 50g feels like, and so on, they will be able to estimate measures effectively. And if you have taught them how to read and create a range of scales, they will be able to transfer this skill into many other areas of mathematics.

If you did not attempt the Portfolio task activities in teaching points 4 and 5, complete them at this point. This will allow you to assess your understanding of the relationship between area and perimeter.

Going further

Progression in measuring

MARGARET BROWN AND HER COLLEAGUES

This report on a research project explores the extent to which learners' progression in learning measurement can be described. As you might expect there are idiosyncratic patterns to individuals' progression in their learning. However, the team is able to outline a general framework to describe progression in this area. It is interesting to compare this to any progression described in other curriculum documents.

Brown, M., Blondel, E., Simon, S. and Black, P. (1995) 'Progression in measuring', *Research Papers in Education*, 10(2), 143–170.

Primary teachers' perceptions of their knowledge and understanding of measurement

MICHELLE O'KEEFE AND JANETTE BOBIS

This paper was presented to the Mathematics Education Research Group of Australia. This study focused on primary teachers' perceptions of their knowledge and understandings of length, area and volume. It also explored their understanding of how children's growth of measurement concepts and processes develops. Data gained from in-depth interviews revealed that teachers' knowledge was often implicit and that they struggled to articulate their knowledge of measurement concepts and children's trajectories of learning.

O'Keefe, M. and Bobis, J. (2008) *Primary Teachers' Perceptions of Their Knowledge and Understanding of Measurement*. Available at www.merga.net.au/documents/RP462008.pdf Accessed 18 July 2012.

Self-audit

These activities focus on estimation of distance, mass and capacity. Through working on the ideas you should develop your understanding of these concepts and be better able to support those you teach.

1   Complete this table. The first column is a measurement; in the second column write down an object that is approximately that length, and in the final column rewrite the measurement in either cm or m, whichever is most appropriate

Length		
1mm		
10mm		
100mm		
1,000mm		
10,000m		
100,000m		

2   Repeat this activity for weight, using g or kg

Weight		
1g		
10g		
100g		
1,000g		
10,000g		
100,000g		

3   And capacity, using ml or l.

Capacity		
1ml		
10ml		
100ml		
1,000ml		
10,000ml		
100,000ml		

CHAPTER 10
STATISTICS AND PROBABILITY

Introduction

Several years ago, Tony taught a class of 10–11-year-old students in his daughter's school in the Midlands of England. The class he was working with included children from a wide variety of backgrounds. Tony was asked to teach statistics to this class and was interested in the group posing their own questions to explore and then to use statistical methods to report in what they had found out. One group suggested carrying out a survey into attitudes towards racism.

The students designed a questionnaire, decided who they would use this questionnaire with and then analysed the responses and reported back to the class. They reported back using **bar charts**, **line graphs** and **pie charts**, with each group describing why they had chosen particular representations to illustrate their data. Some of their results are given in the following table:

How would you feel if someone made fun of your skin colour?

I would hit them	16%
I would be sad	34%
I would be angry	38%
I would think they were ignorant	12%

What would you do if no one would let you play because you looked different to them?

I would ask them why	42%
I would let a teacher know	58%

Two people wrote, "I would be sad and play with someone else," and another wrote, "I would wish I could change my skin colour."

DOI: 10.4324/9781003315155-10

If you were on a bus and you had a spare place next to you, how would you feel if there were lots of people standing and no one would sit next to you?

I would think they didn't like me because I am different   64%
I would ignore it and be happy because I wasn't squashed   36%

This approach to handling data made direct links to the student's lived experience; it also allowed the teacher to bring mathematics directly into the personal, social and intercultural education curriculum that focused on these issues for the next three weeks. It opened the teacher's eyes to issues they hadn't been aware of. The day-to-day prejudice impacted on their students when they travelled on public transport and the damage to self-esteem suffered when children are made to feel that "they want to change their skin colour."

We would suggest that the students came to understand the mathematics of statistics more deeply from this project than they would have done from surveying shoe sizes. It was a question that they were committed to, and they had to decide how to tell the story (or interpret the data) accurately so that people could understand the issue.

Starting point

Let us start with an activity. Work on this task in your notebook.

Portfolio task 10.1

Look at this column graph.

Write down three different sets of data that the graph could be illustrating.

You can define for yourself what the numbers 1–6 along the x-axis (the horizontal axis) stand for.

If you work with friends on this describe how you decided on your answers.

This may seem like starting at the end, but the key reason for using statistics to interpret the world around us is to be able to come to decisions about what the data we collect represents. Often students do not know how to interpret data; they learn how to collect and organise data but do not realise that there may well be alternative interpretations that could be drawn from the data. It is important that our students see the big picture from the beginning and understand that interpretation is just as important as collection and representation of data.

The data handling cycle is shown in the following figure:

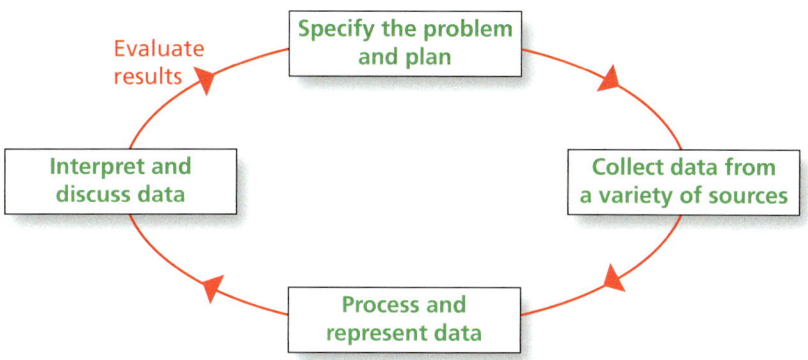

This diagram emphasises the importance of both problem posing and interpretation and is the key big idea for this chapter. Skills in these areas are just as important as understanding how to collect data and the different ways you can represent it.

Progression in statistics and probability

This section illustrates the progression for teaching handling data. The terms in red are defined and explained for you in the glossary.

Foundations for understanding statistics and probability

At the early stages you will work with your students, sorting familiar objects to identify their similarities and differences and counting how many objects share a particular property. They can represent this data using objects and simple **pictographs.** You can encourage them to make simple inferences by asking yes/no questions. You will support the children in presenting their results using pictures, drawings or numerals. As they develop these skills, they will use lists and tables as well as pictures.

Early development of the language of probability will come through talking about events that "will happen" or "won't happen." This will move on to understanding when events are "likely" or "unlikely" and "certain" or "impossible."

Beginning to understand statistics and handling data

At this stage children will be able to ask and research questions by recording information in lists and tables and present outcomes to data handling activities using practical resources, pictures, **column graphs** or **pictographs**. They will begin to use ICT to organise and present data and will use lists, tables and diagrams to sort objects. They will be able to explain the choices they are making.

Students will be able to order the chance of everyday events happening and identify events where if one happens the other cannot.

Becoming confident in statistics and handling data

These students will use a wider range of diagrams to record data including **tally charts**, **frequency tables**, pictographs, column graphs and **dots plots** to represent results and illustrate observations. They will use ICT to create a simple bar chart and use Venn diagrams or Carroll diagrams to sort data and objects using more than one criterion. They will be able to answer a question by identifying what data to collect and how best to organise, present, analyse and interpret the data using tables, diagrams, tally charts, pictographs and bar charts and using ICT where appropriate. They will also be able to compare the impact of representations using different scales. They will understand the difference between **discrete** and **continuous data**.

Students at this stage will be developing the language of probability and the occurrence of familiar events using the language of chance or likelihood. They will understand the idea of **equally likely outcomes** and understand and use the **probability scale** from 0 to 1 and find and justify probabilities based on equally likely outcomes in simple contexts.

Extending learning in statistics and handling data

At this stage you can teach your students how to describe and predict outcomes and to solve problems by collecting, selecting, processing, presenting and interpreting data drawing conclusions in order to identify further questions to ask. They could construct and interpret frequency tables, bar charts with grouped discrete data and line graphs. They will explore hypotheses by planning surveys or experiments to collect small sets of discrete or continuous data, selecting, processing, presenting and interpreting the data in order to identify ways to extend the survey or experiment.

Our aim is that this section illustrates to you the progression in the ideas that help us collect and interpret data. In the Early Years, students are sorting objects and talking about the decisions being made in order to sort according to particular criteria. Then, as they progress, they are introduced to an increasing range of charts and diagrams that can be used to represent data until they are able to explain the results of a survey they have carried out and justify their choices of methods and representation.

Big ideas

The data handling cycle introduced earlier is the big idea when working with students on developing their understanding of statistics. Collecting, organising and interpreting data is at the heart of the mathematical ideas here. Another important way to understand and interpret data is through the key ideas underpinning probability and chance, so these concepts are explored in this section.

Collecting data

Before we begin to collect data, we need a question we are trying to answer. It is important that students are exploring data to answer questions that they want to know the answer to and that they are supported to explore new questions that are raised by the data they have collected. An example from the classroom is the question:

What is the ratio of the circumference of your head to your height?

One class began by estimating answers, guessing from "4-times" upwards. They then start measuring and plot the measurements on a scatter diagram. This means using one axis for height and one axis for circumference of your head.

Each student finds the point on the chart where they would place their data, marks a cross and initials it. By plotting boys and girls in different colours and children and adults in different colours, the students notice that, although the ratio is about 3:1, it may be slightly different for boys and girls and for adults and children. One group decided that they would go into a Foundation class and see if the ratio was different for very young children. Even though scatter graphs may not traditionally be introduced until secondary school, the children were fascinated to see this way of illustrating data.

Organising data

One of the important skills in handling data is deciding what the most appropriate way to organise the data is once it has been collected. This depends on the sort of data you are collecting and the questions you want to answer.

Discrete and continuous data

Discrete data can be counted. Examples would be the ways that students travel to school (five travel by bus) or birthdays (three students have birthdays in May). Continuous data are measured. Examples would be heights, weights and time. Continuous data has to be grouped in order to represent and interpret it. So, for example, if the students wanted to find out how fast their class was compared to other classes, they could devise an activity timing how long it takes individuals to run 25m. In order to represent the data, the times would have to be grouped, perhaps into two-second intervals. The students would need to decide the most appropriate interval after collecting the data. It is worth organising the data using a range of intervals. The students can then decide which is the most useful.

Pictographs, bar graphs, line graphs and pie charts

Initially, you will encourage young students to create pictographs through a practical activity. For example, the children could explore a shop in the role-play area and draw pictures to record their favourite purchases. A question such as:

What do what people in your class like to buy?

would encourage more in-depth interpretation of the data. You could also introduce different roles into the role play; the students could be teachers out shopping or people who work on a building site. They could create different pictographs for these different groups of people.

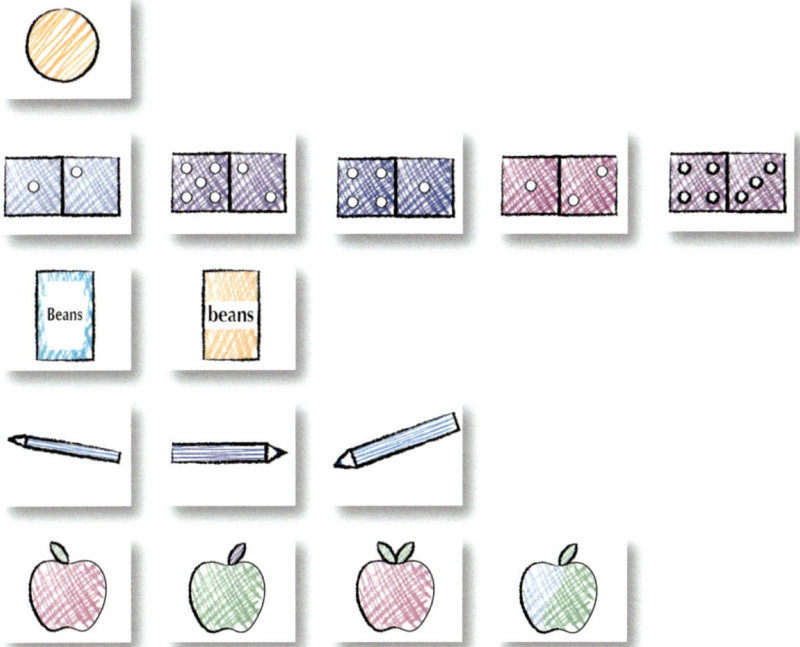

The next activity asks students to compare representations of data using a bar graph and a line graph.

Column graph and dot plots

In this activity you can see two representations of data that encourage the children to explore which representation they think shows the data most usefully.

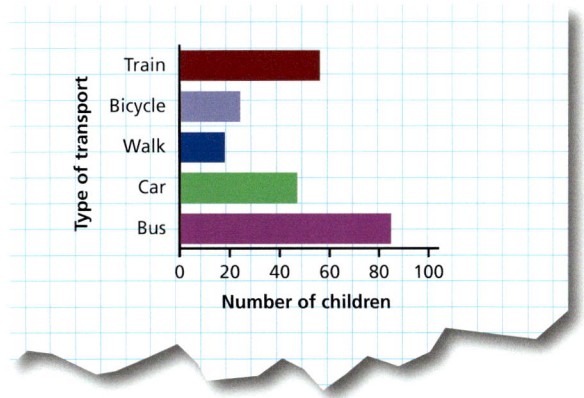

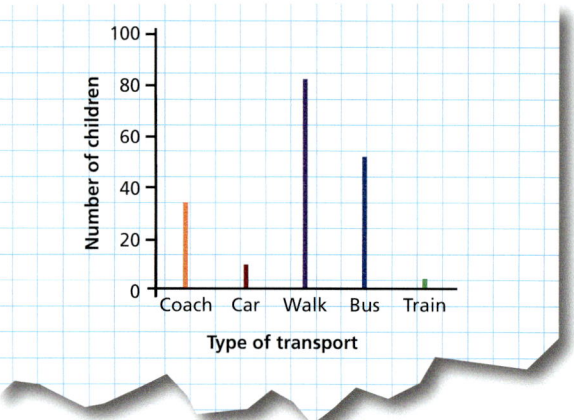

but change data in line graph so the two data sets are identical.

Your role when working with data represented in this way is to use the page for discussion. Ask questions such as:

● which representation is most useful?
● what does the data show – about the children?
● where do you think this school is?

As a follow up to this activity, which focuses on interpreting data, you could collect similar data from your class or from the school and then compare the data for your class with the data for "Bancroft School" and "Parson's School."

Line graphs are used to look at trends over time. You will often see these types of graphs on the news. You can collect and interpret data using line graphs to answer questions like:

● is it too hot/cold in our classroom?
● does the classroom take a long time to warm up in the morning?
● what happens on particularly sunny or very cold days?
● how does a plant grow over time?
● how does the temperature of water in a container change if I move it from the fridge to outside in the sunshine?

The following figure shows the type of axes you would use to record temperature change in the classroom.

Title: _____

Temperature (°C)

30
28
26
24
22
20
18
16
14
12
10
8
6
4
2
0

09:00 09:30 10:00 10:30 11:00 11:30 12:00 12:30 13:00 13:30 14:00 14:30 15:00

Time

Students are also expected to be able to interpret pie charts. Pie charts show the proportions of data. Work on the following task and add it to your portfolio.

Portfolio task 10.2

Examine this pie chart. Write down three different sets of data that it could represent. Choose one of your sets of data and write a short paragraph analysing the data. This question is deliberately open as it exemplifies how data can be interpreted in many different ways.

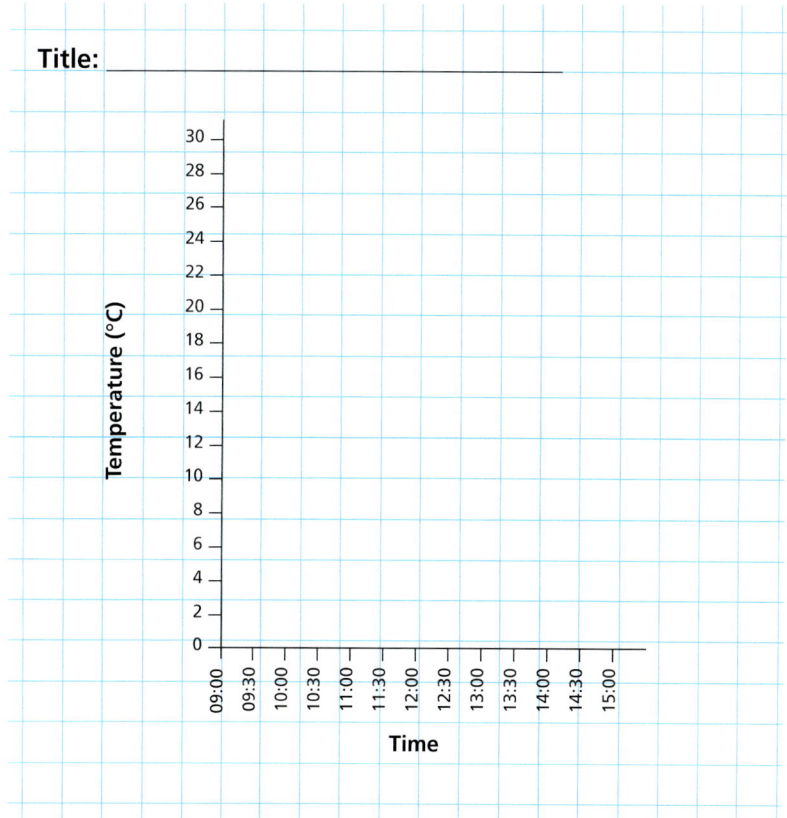

UK 24%
US 20%
Australia 8%
France 16%
Germany 32%

Venn and Carroll diagrams

Venn diagrams and Carroll diagrams are used to sort objects. Carroll diagrams are actually named after the author Lewis Carroll, who wrote *Alice in Wonderland* and was fascinated by mathematics. People have explored *Alice in Wonderland* for the mathematics it contains. Lewis Carroll also wrote academic books on geometry. Venn diagrams were introduced by a mathematician called John Venn in 1880. The task below illustrates the different uses of Carroll and Venn diagrams. In fact, you have already used Carroll diagrams for sorting in Chapter 5.

Portfolio task 10.3

Look at the following shape sorting task and the number sorting activity that follows. Carry out the tasks and then find ways of sorting the shapes using Venn diagrams. What are the advantages of using a Carroll diagram for sorting? What are the advantages of using a Venn diagram for sorting?

Carroll and Venn Diagrams

CARROLL DIAGRAM

Sort the shapes and write the letters in a Carroll diagram like the one in Figure 10.7.

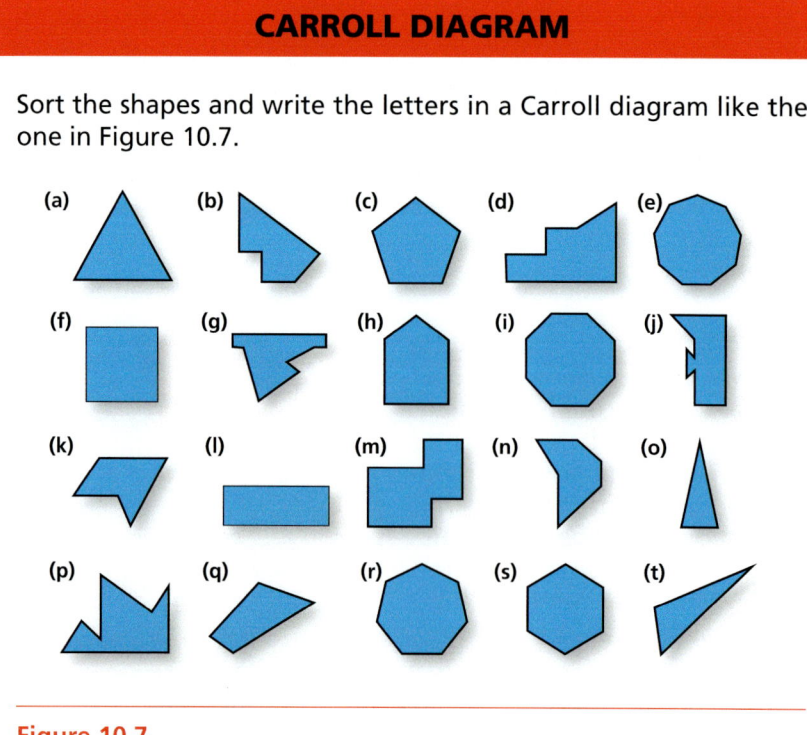

Figure 10.7

	Symmetrical	Not symmetrical
More than four sides		
Four sides or fewer		

Venn diagram

Write the multiples of 6 and 7 on this Venn diagram. The intersection should show the numbers that are multiples of both 6 and 7.

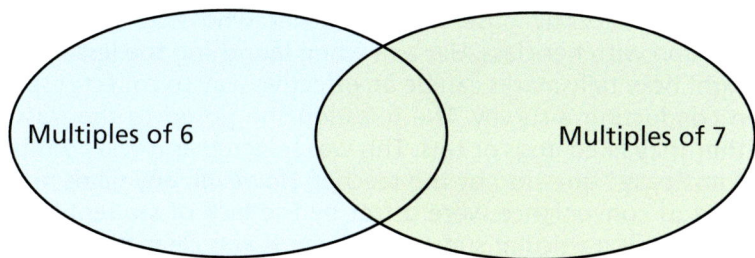

Chance and probability

Children have intuitive ideas about probability through the language that they use. Look at the phrases below and complete each sentence:

It is very likely that . . .

Once in a blue moon . . .

There's a 50/50 chance that . . .

It's unlikely that . . .

It's certain that . . .

It's impossible for . . .

Now organise these phrases on a line so that the least likely to happen is at the left-hand end and the most likely at the right-hand end. You have just created a probability scale. All probability is measured on a scale of 0 to 1 with 0 being impossible and 1 being certain; percentages are sometimes used, with 100% being certain. There are three ways that we can assess the probability of an event happening. First, we draw on prior scientific knowledge. An example of this would be predicting a 70% probability of rain given current weather patterns. Another way of assessing probability is to carry out an experiment. We could work out the likelihood of a piece of toast landing butter-side down by dropping 100 pieces of toast and seeing how many actually landed butter-side down. Finally, we can calculate probability based theoretically on equally likely outcomes. The way we calculate the probability of throwing a 5 on a six-sided dice is by listing the six equally likely outcomes 1, 2, 3, 4, 5 or 6; throwing a 5 is one of these outcomes, so the probability of throwing a 5 is 1/6 (1 out of 6). More formally we write this as $p(5) = 1/6$.

Teaching points

Teaching point 1: making data meaningful

Michael was recently observing a teacher who was introducing a data session with her class. Her aim when launching the lesson was to highlight how tally marks can be an effective way to collect responses when conducting a survey. The question she posed to the class was whether they liked dogs or cats. This was selected as it was considered to be an "easy" question by the teacher. However, any gains made in the area of convenience were offset by the lack of student engagement with this particular survey question. It was clear that many students did not have a strong preference for either dogs or cats, thus they showed very little interest in the task.

The question was posed to the entire class, who were sitting on the floor in front of the whiteboard. She asked those students who liked dogs to stand up and then selected a student to count this group. There were 17 students standing, which the teacher then recorded on the whiteboard as three collections of 5 tally marks, with 2 additional marks. This method of data collection completely removed the need for tallying. In fact, counting the collection and then recording the total of 17 using tally marks was an extremely inefficient way of collecting the data, and this is likely to have left many students unsure of exactly why tally marks are used.

The class did not engage with the data collection in any meaningful way as they were not really interested in the outcome. This is in contrast to the activity that opened the chapter. You will be amazed how much more interest your students will have in learning about statistics if they are exploring a question they are interested in. An area that we have found many students have been interested in has been linked to the environment. One class became very engaged in finding out which class in their school was the "greenest." Together, they developed ways of measuring this characteristic and plotted their findings in meaningful ways to come to a conclusion and communicate their results to others.

TAKING IT FURTHER: **FROM THE CLASSROOM**

In their article "Teaching statistics at the primary school level: Beliefs, affordances and pedagogical subject knowledge," Helen Chick and Robyn Pierce from the University of Melbourne acknowledge that the use of "real-life" data in statistics is important. They designed a series of lessons designed to explore water storage levels and capacity in and around Melbourne, a topic that is becoming increasingly important. They worked with pre-service teachers and asked them to use data they provided to plan lessons for 11-year-old students. They discovered that many of these pre-service teachers did not have the pedagogical subject knowledge to identify what was important in the data and how they could use this data to help them plan effective lessons. For many of these pre-service teachers, concerns with "covering" the curriculum overrode concern with exploring the data critically.

So, there is work for us all to do in better developing our subject knowledge so that we understand that developing a critical understanding of statistics can be done alongside "covering" the curriculum. Indeed, our students are more likely to be able to understand the skills and knowledge they need in statistics if they have been exploring real-world data.

Teaching point 2: making sensible observations from data

This teaching point follows directly from the point made earlier. If students (or indeed teachers) do not see the data as meaningful or have not been involved in formulating the questions that they are trying to respond to through collecting and analysing data, they may well only see simplistic, closed interpretations of data. The interpretation of data should take place alongside the representation of data. In this way children are always making decisions about the most appropriate form of representation. Students should also be encouraged to come up with alternative interpretations of data so that they become used to being critical about the way that data is represented and interpreted.

TAKING IT FURTHER: **FROM THE RESEARCH**

In the paper "Exploring the complexity of the interpretation of media graphs," published in *Research in Mathematics Education* (Volume 6, 2004), Carlos Monteiro and Janet Ainley describe the way they used graphs from the media to develop a "critical sense" in learners' interpretation of data. They suggest that there are three main types of graph reading:

1 reading the data: lifting information from the data to answer specific questions
2 reading between the data: finding relationships and patterns within the data
3 reading beyond the data: using the data to predict future patterns or to ask new questions

The example used earlier in the chapter exploring the relationship between a person's height and the circumference of their head showed learners carrying out all three processes.

The researchers used graphs and charts from the media to explore the ways in which student teachers interpreted data, encouraging them to read the data in the three ways described earlier. They found that the student teachers drew on four aspects of prior knowledge to interpret the graphs:

1 mathematical knowledge to describe the quantitative relationships they observed (that is, relationships based on numbers)
2 personal opinions to make generalisations based on the data
3 personal experience to make generalisations based on the data
4 feelings and emotions to describe how the data made them "feel."

Thinking again about the data handling activity that opened the chapter, you can see an emotional response to the data as it is related to a real, lived experience.

Portfolio task 10.4

Find a chart or graph from a newspaper or magazine which illustrates data that you are interested in. Write down your own analysis of the data. Try to decide when you are "reading the data," "reading between the data" and "reading beyond the data."

Teaching point 3: probability – equally likely outcomes

A boy once gave the best explanation of probability and equally likely outcomes we have ever heard. He was carrying out an experiment tossing a coin and had tossed three heads in a row. His friend said, "The next one is bound to be a tail." The first boy responded, "Don't be silly, the coin hasn't got a memory." And he was right, the probability of tossing a tail is always 1/2 or 50% or 50/50, whatever you have just tossed. Similarly, many students seem to think that it is harder to get a 6 when throwing a dice than any other number. This is not the case. The chance is 1/6, the same as the chance of getting a 1 or a 2 or any other number. Perhaps students just remember times when they could not throw a 6 to start or finish a board game and so think this is more difficult.

The outcomes of an experiment are equally likely to occur when the probability of each outcome is equal. Tossing a head or a tail on a coin or throwing a 1, 2, 3, 4, 5, 6 on a six-sided die are called equally likely outcomes. To find the probability of an event happening you need to decide which equally likely outcomes are acceptable and divide that by how many equally likely outcomes there are. For example:

The probability of throwing an even number on a six-sided dice = 3/6 or 1/2.

There are three possible outcomes that are acceptable, throwing a 2, 4 or 6, and there are six possible outcomes altogether. You can explore theoretical probabilities when events are not equally likely by carrying out a range of experiments. For example, tossing a match box; you can get the whole class tossing (empty) match boxes. If you were to toss it a total of 100 times and it landed on the end 22 times, the side 4 times and the faces 74 times, the probabilities would be 22%, 4% and 74% respectively. You could also explore how the judicious placing of sticky tack changes the probabilities. You could set a challenge to try to reach 50% probability of landing on one end by weighting the box.

In practice

The lesson plan that follows was used to support a group of students in their understanding of data and statistics in a Year 5 and Year 6 multi-age classroom. Following the plan is an evaluation of the lesson that explores how successful the plan was in supporting the students to develop their knowledge, skills and understanding.

Learning intentions

To solve problems by collecting, processing, presenting and interpreting data

To draw conclusions and identify further questions to ask

To suggest, plan and develop lines of enquiry

Key vocabulary

survey, observation, data, hypothesis

Resources

Presentation software e.g., PowerPoint

Context for lesson

This lesson was carried out towards the end of a data and statistics unit that focused on using data to explore real-life questions. I wanted to work on an extended activity with the group so that they could see the data handling cycle as a continuous process. I also wanted the group to see that often a data handling investigation throws up new questions for us to consider.

Warm up activity: observations vs interpretations

As students walk into the classroom in the morning, I sometimes like to have them answer a question to get them engaged with the day ahead and to generally build relationships and wellbeing. I display the question on the whiteboard, and to answer the students slide their mini self-portrait above their answer. Questions range from factual to emotional check-ins to personal preferences, for example, "What did you have for breakfast today?" or "How are you feeling this morning?" or "Which team do you barrack for in the AFL?"

Together we deep-dive into the results, sorting our statements through a T-chart: Observations vs Interpretations. Some of the observations made are simple statements such as "Most people go for Essendon" and "No one in our classroom barracks for Brisbane."

We then dig deeper when we discuss our interpretations by asking why. "Most people in our classroom might go for Essendon because that's the suburb closest to our school" and "Sophie might barrack for Port Adelaide because one of her parents is originally from Adelaide."

This daily routine can easily be a part of any F-6 classroom.

Launch

Write on the whiteboard, "The students in this class enjoy reading." Ask the class how they could prove or disprove whether this statement is true and record their suggestions so that they're visible.

When I carried out this lesson some of the suggestions were: take a simple vote through a survey, record observations of the amount of time students spend reading each day and the reported number of different genres students enjoy reading.

Then give students time in pairs to brainstorm a list of statements about the class that they would like to prove or disprove. After a few minutes share and discuss all of the statements, deciding as a whole group which are the six most interesting hypotheses about the class. The students will choose which group to work in depending on the hypothesis they are most interested in.

Explore

Each group should spend ten minutes deciding how they will test their hypothesis in at least three different ways, then present these criteria to the class for feedback.

For example, the group exploring "The students in this class find Assembly boring" decided to prove their theory by conducting a survey, recording the number of yawns by students during Assembly and observing the number of students fiddling during Assembly.

Once each group has amended their criteria, they begin collecting the data. They do this by creating and carrying out their surveys, organising when the observations will be carried out and deciding how many times they will need to conduct these collections of data to prove or disprove their hypothesis.

Finally, the group decides how best to represent and display their data. Each group will prepare a short PowerPoint presentation to present their findings.

Summary

Each group presents their findings, and the rest of the class provides comments through the two stars (two positive features) and one wish (one way of improving it) system of feedback.

Rationale and evaluation

This project-type task needs flexibility to manage the required time differently. In fact, I had to build in some flexible time during the week for groups to finish their presentations as they took very different time scales. I should have expected this, but it was important that the groups could take the time they needed over preparing their presentations as it allowed them to draft and redraft. I asked each group to show me their data as pie charts, bar graphs and line graphs if they could. I then asked them to make the decision about which was the most useful representation. I think this was the first time that they thought carefully about how pie charts are better at representing "proportion" than bar graphs. Charlotte said, "I'm going to use a bar chart for the information that has lots of numbers that are important but the pie chart when I just need to show where the most is." I thought this showed a good understanding of the alternative representations.

All the groups found the hardest part deciding on the initial criteria. I needed to support them carefully initially as all the groups wanted to use just one criterion. For example, they asked, "Do you like to read?" rather than, "Do you read action stories, or poetry books, or science books?" In the end, I had to structure the learning by asking for three criteria they could explore. However, when we revisited the criteria at the end and started thinking of new questions to ask, they could see how they might devise criteria. There has been some impact I think, as yesterday I said, "Well done Room 17, you've worked very hard today." Liam chirped up, "What do you mean exactly by we've worked hard?"

Having mixed-ability groups seemed effective; they allowed people like Paul, who is normally very quiet, to take a lead. He became very animated as he worked on the question, "Children in this class are good at looking after animals," as that is his passion. Similarly, Miranda and Angus became the IT experts for their groups and showed them all sorts of techniques for using presentation software.

The groups' skills in peer assessment were also developed. They are becoming much more sophisticated in using the two stars and one wish feedback system. Earlier in the year the responses would have been along the lines of, "It was very colourful" or "It could have been neater." This time there were comments such as, "I think a pie chart would have been a better way of showing the data because you could very quickly see the different fractions."

Assessing knowledge of statistics and probability

It is important that the activities you use to assess data handling treat the data handling cycle holistically whilst allowing you to assess the component parts. Many of the activities that you have been introduced to in this chapter can be used as assessment activities. For example, the activity presented earlier in the chapter can be used to assess student's abilities in classifying data and explaining the different choices available to them when sorting data. Most importantly it allows you to assess your students' ability to explain the choices they are making. Similarly, the activity presented in the bar and column charts activity, if extended into an activity that compares this data with data collected about your class, allows you to assess across the data handling cycle.

Cross-curricular teaching of statistics and probability

A mathematics project Tony remembers as being very effective in exploring and teaching statistics was setting up a healthy canteen or a school. In order to get the project off the ground, the students had to carry out a large-scale data handling activity. They were asked to decide what information they would need to find out if we were setting up the canteen. This included:

- what do we mean by a "healthy" food?
- what healthy foods would you buy?
- how much would the students in school spend at a healthy canteen?
- where would the best site for the healthy canteen be?
- when would be the best time to run the canteen?
- how much would it cost to buy the stock?

The students opted to join a group they were interested in and then carried out the research necessary to answer the question. They then had to present their findings to the school council as it had the final decision about funding the canteen. The research was a success – the canteen was funded and ran successfully for several years.

Summary

The main focus of this chapter has been the data handling cycle reproduced here:

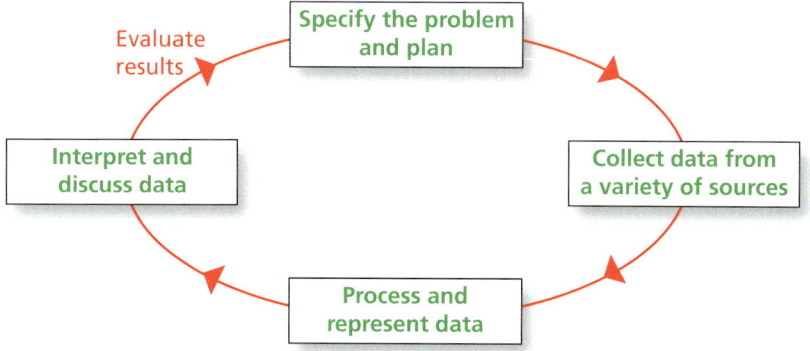

You should have a clearer understanding of each part of the cycle and how important it is to see the cycle as a whole and that working

with students to see the cycle as a whole is central to learning how to handle and interpret data. You have also seen the range of ways in which data can be represented.

Reflections on this chapter

Statistics and data handling can be an exciting and motivating area to explore with the students you teach. It is also, arguably, one of the most important areas of mathematics in which to develop skills in order to be able to make sense of and understand the world around us as presented through the media, a skill becoming greatly valued by employers. We would hope that by gaining confidence in seeing the data handling as a process of posing questions, gathering data that is then interpreted to answer those questions and finally using this process to ask new questions, you will have become excited at the prospect of working through this cycle with your students.

It is possibly the only area of mathematics that can be taught through a totally cross-curricular approach. Indeed, it can possibly only be taught in this way. Similarly, it is an area of mathematics that is best explored by drawing on the students' interests as a starting point.

Several years ago, Tony was having a drink with a cousin of his whom he didn't see very often; out of the blue she asked him if she should stop taking the contraceptive pill. He was taken aback and asked why she wanted to know. She told him that she had read in the paper that being on the pill doubled her chance of getting thrombosis in her legs. He asked her to show him the article and was able to explain to her that even doubling the risk still meant that the risk of thrombosis was very small. He realised she had asked him because she saw him as "good at maths," in her words. She didn't feel confident in making important decisions based on her data handling skills.

In recent times we have been presented with a wealth of statistical data to try and explain the risks that the COVID-19 virus presented. People have been asked to make major changes to their lives and to adapt behaviours based on interpretations of this data. It has not always been clear that our politicians have a good grasp of the mathematics that underpins the statistics we are presented with.

So, we would argue that teaching statistics and data handling is very important. Enjoy it!

Going further

If the world were a village

DAVID SMITH AND SHELAGH ARMSTRONG

This invaluable resource allows you to work with your students on issues of global importance whilst introducing them to ways of representing and interpreting data. The book uses the construct of imagining the world was a village of 100 people to present data around issues such as access to electricity, the number of languages spoken globally and poverty.

Smith, D. and Armstrong, S. (2004) *If the World Were a Village: Imagine 100 People Live in a Village*. London: A & C Black.

Teaching statistics at the primary school level: beliefs, affordances and pedagogical content knowledge

HELEN CHICK AND ROBYN PIERCE

This article, discussed in the previous chapter, explores the importance of teacher subject knowledge in the planning and teaching of statistics. Particularly of we are trying to engage students with real world issues.

Chick, H.L. and Pierce, R.U. *Teaching Statistics at the Primary School Level: Beliefs, Affordances and Pedagogical Content Knowledge*. Available at https://iase-web.org/documents/papers/rt2008/T2P3_Chick.pdf Accessed 17 June 2021.

Data handling

J MICHAEL SHAUGNESSY, JOAN GARFIELD AND BRIAN GREER

These authors have contributed a chapter on data handling in the most recent *International Handbook of Mathematics Education*. This explores the historical role of data handling in mathematics curricula around the world and pays particular attention to the way in which ICT can support the learning and teaching of data handling.

Shaughnessy, J.M., Garfield, J. and Greer, B. (1996) *The International Handbook of Mathematics Education*. Bishop, A. (ed.). Dordrecht: Kluwer Academic Publishers.

There are also a range of websites that are absolutely fantastic resources for exploring global data:

www.gapminder.org/

www.dear-data.com/theproject

https://informationisbeautiful.net/

Self-audit

Gather some achievement data that you have on a class that you are teaching. This may be previous achievement on NAPLAN, it may be results using Running Records in reading – it can be any quantitative data. Decide what you want to find out from the data, such as comparing boys' and girls' achievements or a particular subject area you have been focusing on, and write a list of questions. Use the data handling techniques you have met in this chapter to analyse the data. Also use a range of ways of representing the data. Use your analysis to write a short report on the achievements of your class. Include this report in your portfolio.

CHAPTER 11

TEACHING AND LEARNING MATHEMATICS IN THE EARLY YEARS

Introduction

This chapter follows a slightly different pattern from previous chapters to acknowledge that experiencing mathematics for children in the Early Years might feel different from learning mathematics later in primary education. Specifically, children should be much more involved in initiating their own learning and should be developing their mathematical skills through activity. However, of course we would hope that there will be plenty of activity and student-centred learning throughout the primary school.

Tony tells the story of being out for a walk with his grandson when he saw a truck. Felix could see three wheels on one side of the truck. He said to Tony, without any prompting, "That truck has got six wheels." He did not expand on this, just stated it as a fact. He did not want to explain any further about how he knew this and avoided continuing the conversation when Tony pointed out a truck with four wheels on one side. This story suggests an alternative view on the notion of progression with which we have opened each chapter with so far.

The booklet *Developmental milestones and the Early Years Learning Framework and the National Quality Standards*, which outlines expectations for children in Early Years settings in Australia, opens with a quote from page 19 of the *Early Years Learning Framework*. This states:

> Children's learning is ongoing and each child will progress towards the outcomes in different and equally meaningful ways. Learning is not always predictable and linear. Educators plan with each child and the outcomes in mind.

Let's explore this a little further with another story from Tony's grandchildren. Every Saturday morning he goes to a nearby city on the train with his grandson, Tate, aged four when we wrote this section. Tate has started noticing numbers all around him. On the last trip he saw a sign indicating the gradient with a "10" on it and said, "a one and a zero, that makes ten." Later he saw another sign, "What does a four and a zero make Grandad?" "Forty," Tony suggested, and Tate seemed content. Tony felt in no rush to explore place value, plenty of time for that. What excited him was Tate's propensity for noticing numbers, for looking at patterns and in this case generalising. A single digit and a zero "makes" one of the "tens numbers."

Next, they counted numbers of people getting on and off at each stop, an activity Tate initiated. He wanted to show the numbers using his fingers. "I know that one, its seven," he said making it this way.

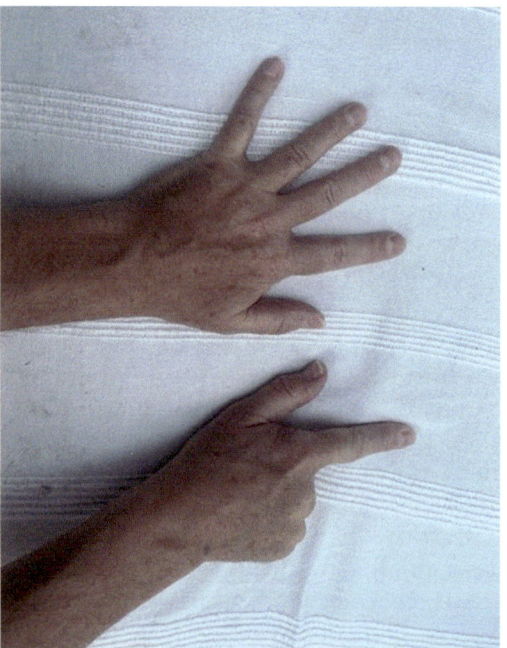

Or at least tried to! It's quite hard to tuck the three fingers under – try it. Tony then showed him this way and suggested that it was "seven" as well.

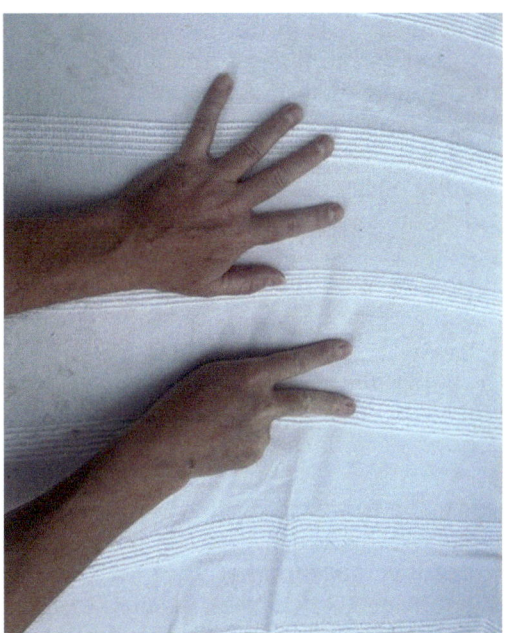

Tate laughed, "No – that's not right," he said. Tony asked him to count the fingers and he was surprised to find there were seven. So surprised that he counted them again, and then deliberately mis-counted to count eight. And as for this way:

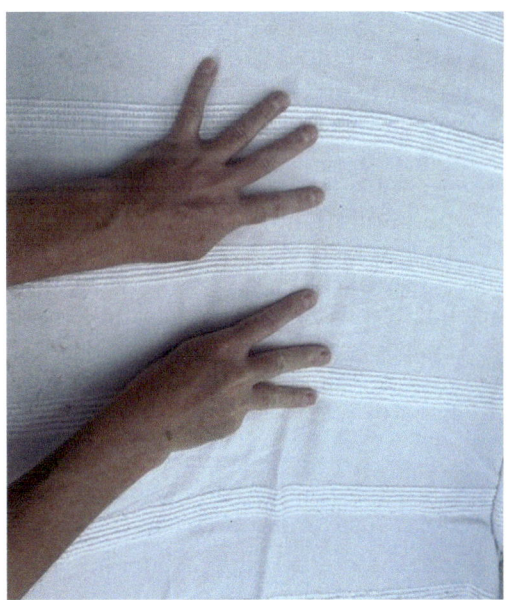

Well, it doesn't even use either thumb, and it doesn't use all five fingers on either hand. Plainly ridiculous, even though careful counting, twice, seemed to suggest there were seven fingers in the representation. We wonder if this was the first sign of school mathematics clashing with the real world of mathematics. One guess might be that, in his Early Years setting, Tate had been being "taught" number pairs and had been shown a single finger pattern for each number so he could write 5 + 1 = 6; 5 + 2 = 7; 5 + 3 = 8 and so on. Maybe his setting felt pressured to meet the new Early Learning Goal in England:

> *Automatically recall (without reference to rhymes, counting or other aids) number bonds up to 5 . . . and some number bonds to 10, including double facts.*

So, a unique landscape to be explored as an individual has become a journey experienced at the same rate as his peers. But what is this progression? Tony recently adapted a series of textbooks from Singapore to follow the Cambridge curriculum. The curricula interpret progression differently. In the Cambridge curriculum fractions are introduced gradually, in the Singapore curriculum all unit fractions up to $\frac{1}{10}$ are introduced simultaneously. This begs the question, "Why is the notion of $\frac{1}{2}$ more difficult than that of $\frac{1}{7}$?"

We write this to emphasise that progression in curriculum documents is a sequence that helps us plan, and this is what you have been presented with so far, not a sequence that has been specifically designed supports students' learning. Sometimes our governments or administrators seem to expect learning to progress in broadly the same way for all students. This directly contradicts the earlier statement about learning nor being linear. Education in mathematics is when unique individuals develop and learn according to their needs and interests. In the Early Years we need to try to balance these two approaches. As Early Years educators we should also remember that we should be presenting mathematics in an integrated way. *Belonging, Being & Becoming – The Early Years Learning Framework for Australia* details five learning outcomes from birth to five years.

Children have a strong sense of identity:

- children feel safe, secure, and supported
- children develop their emerging autonomy, inter-dependence, resilience and sense of agency
- children develop knowledgeable and confident self-identities
- children learn to interact in relation to others with care, empathy and respect.

Children are connected with and contribute to their world:

- children develop a sense of belonging to groups and communities and an understanding of the reciprocal rights and responsibilities necessary for active community participation
- children respond to diversity with respect
- children become aware of fairness
- children become socially responsible and show respect for the environment.

Children have a strong sense of wellbeing:

- children become strong in their social and emotional wellbeing
- children take increasing responsibility for their own health and physical wellbeing.

Children are confident and involved learners

- children develop dispositions for learning such as curiosity, cooperation, confidence, creativity, commitment, enthusiasm, persistence, imagination and reflexivity
- children develop a range of skills and processes such as problem solving, inquiry, experimentation, hypothesising, researching and investigating
- children transfer and adapt what they have learned from one context to another
- children resource their own learning through connecting with people, place, technologies and natural and processed materials.

Children are effective communicators:

- children interact verbally and non-verbally with others for a range of purposes
- children engage with a range of texts and gain meaning from these texts
- children express ideas and make meaning using a range of media
- children begin to understand how symbols and pattern systems work
- children use information and communication technologies to access information, investigate ideas and represent their thinking.

We also think it is worth reminding you all of the vision in which these outcomes are embedded. This vision can be seen to apply to education at any age. And we should ask ourselves how we model these principles in the teaching of mathematics:

Belonging: experiencing *belonging* – knowing where and with whom you belong – is integral to human existence. Children belong first to a family, a cultural group, a neighbourhood and a wider community. *Belonging* acknowledges children's interdependence with others and the basis of relationships in defining identities. In early childhood and throughout life, relationships are crucial to a sense of *belonging*. *Belonging* is central to *being* and *becoming* in that it shapes who children are and who they can become.

Being: childhood is a time to be, to seek and make meaning of the world.

Being recognises the significance of the here and now in children's lives. It is about the present and them knowing themselves, building and maintaining relationships with others, engaging with life's joys and complexities and meeting challenges in everyday life. The early childhood years are not solely preparation for the future but are also about the present.

Becoming: children's identities, knowledge, understandings, capacities, skills and relationships change during childhood. They are shaped by many different events and circumstances. *Becoming* reflects this process of rapid and significant change that occurs in the Early Years as young children learn and grow. It emphasises learning to participate fully and actively in society.

As we begin to plan mathematical experiences for young learners, we should ask ourselves how these activities are contributing to our young learners' overall experience of the curriculum.

TAKING IT FURTHER: **FROM THE RESEARCH**

In her article, *Supporting early mathematics learning in early childhood settings*, Marianne Knaus from Edith Cowan University argues that many Early Years educators, who are skilled in planning engaging and active learning experiences for young learners, do not always see how these activities can be used to develop mathematical skills. This is often due to their own educational experiences in terms of mathematics and a deficit self-identity of themselves with respect to mathematics. She says:

educators explained that their experiences were provided in a play context including water play, cooking, using play-dough, measuring, puzzles, computer games, songs and rhymes. However, many participants struggled to know what mathematics concepts they were teaching when introducing these experiences.

We hope that the rest of this chapter supports all Early Years educators in seeing the mathematics in the activities they plan. We also think this provides a compelling argument as to why Early Years educators may find the rest of the book useful!

Starting point

Portfolio task 11.1

Explore this with a group of friends. One of you should read out this list of numbers one at a time. After each number is read out draw the images that you see or associate with the number.

8; 25; 6; 13; 100; 1 000 000

You will draw a range of images. Discuss the reasons that lie behind your choice of image. These are often influenced by life experiences. For example, people may see boxes of eggs for 6:

An egg box is a useful image. It means that as well as seeing 6 we see 6 as 2 × 3 or 3 × 2; we may also see 5 as an egg box with one missing, so we know that 6 − 1 = 5. I have seen schools beginning to use egg boxes with 10 "holes" to model counting to 10. Of course, two of these boxes takes us up to 20.

This sort of an activity develops subitising skills. *Subitising*, a term coined by Piaget, is the ability to see how many objects there are in a set without counting them. This supports children in developing a feel for number and allows for the development of mental methods later.

As numbers get bigger, it becomes harder to subitise. Our images become more abstract, or our images don't help us relate to the size of a number. Often we may hear children referring to any big numbers as "millions and trillions and millions." This just really means "big." So, children's early mathematical experiences are very important in providing them with images that they will carry with them as they grow.

Harry, a friend's son, was three when he was shared a book called *Window* by Jeannie Baker with Tony. This is a beautiful book, the winner of the Young Australian's best book award, consisting of drawings of the view through a window on each birthday of a child from ages 1 to 20. In each picture the birthday card is displayed so that the numeral can be seen. As they flicked through the book Harry made no comment until we reached the child's fifteenth birthday. Harry looked at the card and said, "I know that one – that's 15. Sam is 15." Sam, Tony's youngest son, and at that time a hero of Harry's, had just had his fifteenth birthday. Tony remembered that Harry had asked him to show him "how to write 15" when he was told that Sam was 15 years old. So, even though Harry was not writing numbers in order, he was recognising numbers that were important to him and beginning to order and describe his world using numbers.

Similarly, Tony's two sons were out with their grandad when they were very small and walking down our road. One of them started chanting 2, 4, 6, 8, 10, 12 until he reached our door and said, "we live at 14." Tony had, of course, at an early age made sure the boys knew their address in case they got lost. The other one picked up the chant using the houses on the other side of the road, counting 1, 3, 5, 7, 9, 11. Here we can use evidence of young children noticing pattern in number and using these patterns to describe and navigate their world. There is a view from some educators that it is best to limit the numbers that young children are exposed to to those below 20. However, we must be careful not to restrict children to numbers that we feel are within their understanding. Many children live at house numbers larger than 20, have seen three-digit numbers on the fronts of buses and on car registration plates and may have a birthday on the thirty-first! They will see and use numbers larger than 20 in their everyday experience. We must make sure that we draw on this everyday experience rather than make a false dichotomy between "school mathematics and the kind of mathematics that all learners use to understand and interpret their everyday worlds.

Children know bus numbers, know the age of their siblings, understand how to share, may have a sense of money and have a sense of size from very young. We cannot and should not assume that we are working with blank slates when we start exploring number or shape or handling data with our young learners. They have been using mathematics since they were born. It is an exciting task to discover how they are using their mathematics to understand and describe their world. This chapter will explore how every child brings with them their own mathematical understanding and is capable of developing their understandings, with a teacher, who may be an Early Years educator, another adult, a parent or a friend.

Problem solving, reasoning and numeracy

In Chapter 3 we explored how people come to think mathematically. We have listed their key assumptions here to save you flicking back through the book. These were:

- everyone can think mathematically
- mathematical thinking can be improved by "practising reflection"
- mathematical thinking is provoked by contradiction, tension and surprise
- mathematical thinking is supported by an atmosphere of questioning, challenging and reflecting
- mathematical thinking helps in understanding oneself and the world.

We would argue that all these statements can be observed in an Early Years setting. Young children think mathematically whenever they try to impose order on their world; whenever they notice a pattern or whenever they compare objects, and all young learners carry out these activities.

Young students practise reflection through repetition and through the questions they ask. Through this repetition they construct schemas; these are building blocks of knowledge and they use these building blocks to create mental models of the world. You may have noticed a child repeating an action to see if the outcome remains the same. For example, they open a door and notice that the draught from the door means that a balloon blows across the floor. They smile, close the door, move the balloon back to the door and try again. The same thing happens. They repeat again. The next morning, they may try again just to check. They now have a building block, an understanding of a cause and effect.

The third and fourth bullet points give advice as to how we can construct an ethos that will support mathematical thinking and problem-solving. We can create a space in which learners come to expect "contradiction, tension and surprise," and we can support them in developing their mathematical thinking by creating an atmosphere that is questioning and challenging.

You were introduced to *specialising* and *generalising* in Chapter 3 too. At first young learners, in common with many older students, take a special case as true for all cases. For example, if you were to ask your learners to "draw a triangle," I would be almost certain that they would draw an equilateral triangle. We see the special case, a triangle with all sides and angles the same size, rather than the bigger picture. In general, any three-sided shape is a triangle. Much better to give learners many different types of triangles and ask what is the same about all of these shapes

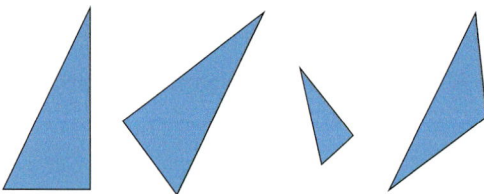

In an article available on the NRich website based at Cambridge University entitled *Mathematical problem solving in the Early Years: Developing opportunities, strategies and confidence*, Sue Gifford describes how Early Years practitioners can support young learners in becoming confident mathematical problem solvers. She suggests that for the youngest learners the stages of problem solving include:

- finding a way to get started – making sense of the problem for myself
- working on the problem
- finding out there is more to the problem than I first thought and exploring further
- deciding I have finished and sharing my solution.

If you reflect on any problem that students in your care have approached, you will recognise these stages. Indeed, if you reflect on your own approach to solving a problem you are interested in you will recognise these behaviours in yourself. However, if the problem our students are presented with is not something they are interested in, they will probably simply rush to a solution to get the problem out of the way so they can spend time working at something that does interest them. So, a huge part of the skill of the Early Years educator is planning activities that will motivate the students (although hopefully this is true for all teachers, regardless of the age of their students) and being flexible enough to know when not to push young children to complete activities if they are not motivated by the problem.

The same article alerts us to the increasing sophistication observable in two- and three-year-old children nesting cups and in children aged four to seven years old building a train track. These stages are described as:

- brute force: pushing the cups down on each other to try to make them fit or hammering train track pieces together to force them together
- local correction: adjusting one cup or one piece of track, which may lead to new problems elsewhere in the stick or in the train track
- dismantling: realising that the best tactic is to start all over again
- holistic review: looking at the problem again and using previous mistakes to take a new approach.

Our guess is that you will recognise this list from observing students you have worked with. We can also think of a few adults who try to apply the "brute force" technique at times! An educator's role in supporting young learners develop their strategic approach to problem-solving could include asking appropriate questions such as:

- what are you trying to do? What problem are you trying to solve?
- have you done anything like this before?
- what do you think you need to solve this problem?
- why isn't that working?
- what can you try instead?
- is that better?
- does it work now? How can we make it even better?
- can you tell your friends how you succeeded?

Young children learning number

Tony tells us many stories about his grandsons and as we were writing this section, he related a counting story from the previous weekend. Tate, aged three, was sitting with him on a train and initiated a counting game, counting up to 100. The game was that Tony and Tate would count alternate numbers aloud. Tate was a little insecure about the multiples of ten, so Tony made sure that they fell on his "turn." After 3 runs through to 100, Tony continued saying "One hundred and one." Tate responded with, "One hundred and two. This carried on until Tony said, "One hundred and nine." Tate responded with, "Thirty." And the process started again with Tony replying "Thirty-one."

This is not to suggest that Tate has a secure understanding of place value, but he is recognising patterns and is certainly able to count beyond 20 making his own interpretation of what the pattern of counting numbers might sound like. So, what might the expectation be?

The Foundation Year content descriptions for number suggest that by the time he moves into primary school he should be able to:

> *Connect number names, numerals and quantities, including zero, initially up to 10 and then beyond.*

So, he is certainly meeting this aim.

The early development of young learners is broken down into stages in the document, *Development milestones and the Early Years Learning framework and the National Quality Standards.* The document reminds us that educators should "use this reference as a source of information rather than a prescriptive checklist." We are interested that the document chose not to specify mathematics-specific outcomes and wonder why that might be. If the milestones include language and early reading milestones, it might be helpful to Early Years educators to have some guidance about early mathematical development. For this reason, the following table includes both the Australian milestones and those from similar guidance in England.

It includes suggested things that adults might do to support this development.

Age range	Development milestones: Australia	Development milestones: England	What adults could do
Birth–11 months	Learns through sensory experiences. Repeats and copies sounds.	Notices changes in number of objects/ images or sounds in group of up to three.	Sing number rhymes as you dress or change babies. Move with babies to the rhythm patterns in familiar songs and rhymes.
8–20 months	Transfers objects from hand-to-hand. Pokes and picks up small objects. Imitates sounds and actions. Enjoys finger rhymes. Imitates hand clapping.	Develops an awareness of number names through their enjoyment of action rhymes and songs that relate to their experience of numbers. Has some understanding that things exist, even when out of sight.	Encourage babies to join in tapping and clapping along to simple rhythms.

Age range	Development milestones: Australia	Development milestones: England	What adults could do
16–26 months	Scribbles with pencil or crayon held in fist. Curious and energetic. Points to objects when named. Comprehends and follows simple commands. Enjoys rhymes and songs. Stacks and knocks over objects. Mimics household activities.	Knows that things exist, even when out of sight. Beginning to organise and categorise objects, e.g., putting all the teddy bears together or teddies and cars in separate piles. Says some counting words randomly.	Use number words in meaningful contexts, e.g., "Here is your other mitten. Now we have two." Talk to young children about "lots" and "few" as they play. Talk about young children's choices and, where appropriate, demonstrate how counting helps us to find out how many. Talk about the maths in everyday situations, e.g., doing up a coat, one hole for each button. Tell parents about all the ways children learn about numbers in your setting. Have interpreter support or translated materials to support children and families learning English as an additional language.
22–36 months	Uses pencil to draw or scribble in circles and lines. Builds tower of five to seven objects. Begins to count with numbers. Can follow two or more directions. Uses correct form . . . of language. Becomes aware of space through physical activity. Recognises similarities and differences.	Selects a small number of objects from a group when asked, for example, "please give me one," "please give me two." Recites some number names in sequence. Creates and experiments with symbols and marks representing ideas of number. Begins to make comparisons between quantities. Uses some language of quantities, such as "more" and "a lot." Knows that a group of things changes quantity when something is added or taken away.	Encourage parents of children learning English as an additional language to talk in their home language about quantities and numbers. Sing counting songs and rhymes that help to develop children's understanding of number. Play games that relate to number order, addition and subtraction, such as hopscotch and skittles and target games. Play games that involve small quantities. For example, shops where you order two cups of coffee or three cakes.

Age range	Development milestones: Australia	Development milestones: England	What adults could do
36–60 months	Imitates variety of shapes in drawing. Independently cuts paper with scissors. Understands opposites. Uses objects to construct things. Builds tower of eight to ten blocks. Counts five to ten things. Follows simple rules. May write some numbers and letters. Counts by rote having memorised numbers. Touches objects to count – starting to understand relationship between numbers and objects.	Uses some number names and number language spontaneously. Uses some number names accurately in play. Recites numbers in order to ten. Knows that numbers identify how many objects are in a set. Beginning to represent numbers using fingers, marks on paper or pictures. Sometimes matches numeral and quantity correctly. Shows curiosity about numbers by offering comments or asking questions. Compares two groups of objects, saying when they have the same number. Shows an interest in number problems. Separates a group of three or four objects in different ways, beginning to recognise that the total is still the same. Shows an interest in numerals in the environment. Shows an interest in representing numbers. Realises not only objects but anything can be counted, including steps, claps or jumps.	Use number language, e.g., "one," "two," "three," "lots," "fewer," "hundreds," "how many?" and "count" in a variety of situations. Support children's developing understanding of abstraction by counting things that are not objects, such as hops, jumps, clicks or claps. Model counting of objects in a random layout, showing the result is always the same as long as each object is only counted once. Model and encourage use of mathematical language e.g., asking questions such as "How many saucepans will fit on the shelf?" Help children to understand that one thing can be shared by number of pieces, e.g., a pizza. As you read number stories or rhymes, ask, e.g., "When one more frog jumps in, how many will there be in the pool altogether?" Use pictures and objects to illustrate counting songs, rhymes and number stories. Encourage children to use mark-making to support their thinking about numbers and simple problems. Talk with children about the strategies they are using, e.g., to work out a solution to a simple problem by using fingers or counting aloud.

Age range	Development milestones: Australia	Development milestones: England	What adults could do
		In UK 50–60 months Recognises some numerals of personal significance.	Encourage estimation, e.g. estimate how many sandwiches to make for the picnic or how many cupcakes are on the tray.
		Recognises numerals 1 to 5.	Encourage use of mathematical language, e.g. "Have you got enough to give me three?"
		Counts up to three or four objects by saying one number name for each item.	Ensure that children are involved in making displays, e.g., making their own pictograms of lunch choices. Develop this as a 3D representation using bricks and discuss the most popular choices.
		Counts actions or objects that cannot be moved.	
		Counts objects to ten and beginning to count beyond ten.	Add numerals to all areas of learning and development, e.g., to a display of a favourite story.
		Counts out up to six objects from a larger group.	Make books about numbers that have meaning for the child such as favourite numbers, birth dates or telephone numbers.
		Selects the correct numeral to represent one to five, then one to ten objects.	Use rhymes, songs and stories involving counting on and counting back in ones, twos, fives and tens.
		Counts an irregular arrangement of up to ten objects.	Emphasise the empty set and introduce the concept of nothing or zero.
		Estimates how many objects they can see and checks by counting them.	Show interest in how children solve problems and value their different solutions.
		Uses the language of "more" and "fewer" to compare two sets of objects.	Make sure children are secure about the order of numbers before asking what comes after or before each number.
		Finds the total number of items in two groups by counting all of them.	Discuss with children how problems relate to others they have met and their different solutions.
		Says the number that is one more than a given number.	Talk about the methods children use to answer a problem they have posed, e.g., "Get one more, and then we will both have two."
		Finds one more or one less from a group of up to five objects, then ten objects.	Encourage children to make up their own story problems for other children to solve.

Age range	Development milestones: Australia	Development milestones: England	What adults could do
		In practical activities and discussion, beginning to use the vocabulary involved in adding and subtracting. Records, using marks that they can interpret and explain. Begins to identify own mathematical problems based on own interests and fascinations.	Encourage children to extend problems, e.g., "Suppose there were three people to share the bricks between instead of two." Use mathematical vocabulary and demonstrate methods of recording, using standard notation where appropriate. Give children learning English as additional language opportunities to work in their home language to ensure accurate understanding of concepts.

We think it is worth reading through this table and noting which activities you already have planned for as a matter of course, or, if you are a carer for a young person, which activities do you engage in during the day simply regarding them as "what you do." Now note which of the suggestions for "what adults do" you do not do. Why do you think you do not do these things? Is it your own nervousness about mathematics, or had you just not thought about the mathematical possibilities for day-to-day activity? We hope you would agree that you can expand your repertoire of everyday mathematical activity to support the developing mathematical understanding of the young learners in your care. And we hope that you find the more detailed mathematical advice from England helpful in seeing how the more generic developmental milestones can be interpreted for early mathematical development. The rest of this chapter adapts and develops these milestones to offer guidance and support for Early Years educators across the different yet connected areas of mathematics.

Young children learning measures

Those of you who have spent time with very young people will know that they like to build towers; to be honest many of us still enjoy building towers. When Tony's grandson, Felix, was two years old he would stand on a small stool and Tony's task was to hand him large connecting blocks so that he could build a tower that was taller than he was. Tony would ask, "Can you reach?" Felix would smile and nod reaching ever higher. Then he would say, "You have to do the next one Grandad – it's too reachy." Young children develop a sense of measurement as comparison well before we see them in our Early Years setting. They understand which object might appear longer (although they do not always match up the ends!) They know that they can lift some things and other things are too heavy. They understand when one object will fit inside another. And they will have learned all this through exploration and activity. The Foundation Year content description suggests:

Use direct and indirect comparisons to decide which is longer, heavier or holds more, and explain reasoning in everyday language.

Our job as educators, then, is to continue to plan activities that will allow our young learners to explore measures. There will, of course, be children who are not fortunate enough to have had these experiences at home; our task here is to ensure all children experience activities involving measurement in the same spirit of activity and discovery. The following table offers more specific ideas. We have adapted the table from the experiences suggested in both the Australian and English milestones documents.

Age range	What a child is learning	What adults could do
Birth–11 months	Babies' early awareness of measure grows from their sensory awareness and opportunities to play and explore.	Provide a range of objects of various textures and weights in treasure baskets to excite and encourage babies' interests.
8–20 months	Recognises big things and small things in meaningful contexts. Gets to know and enjoy daily routines, such as getting-up time, mealtimes, nappy time and bedtime.	Play games that involve curling and stretching, popping up and bobbing down. Encourage babies' explorations of the characteristics of objects, e.g., by rolling a ball to them.
16–26 months	Enjoys filling and emptying containers. Associates a sequence of actions with daily routines. Beginning to understand that things might happen "now."	Talk to children, as they play with water or sand, to encourage them to think about when something is full, empty or holds more. Help young children to create different arrangements in the layout of road and rail tracks. Highlight patterns in daily activities and routines.
22–36 months	Begins to categorise objects according to properties such as shape or size. Begins to use the language of size. Understands some talk about immediate past and future, e.g., "before," "later" or "soon." Anticipates specific time-based events such as mealtimes or home time.	Use descriptive words like "big" and "little" in everyday play situations and through books and stories.
30–50 months	Uses positional language.	Demonstrate the language for measures in discussions, e.g., "long," "longer," "longest," "short," "shorter," "shortest," "heavy," "light," "full" and "empty." Find out and use equivalent terms for these in home languages. Encourage children to talk about the shapes.

Age range	What a child is learning	What adults could do
40–60+ months	Can describe their relative position such as "behind" or "next to." Orders two or three items by length or height. Orders two items by weight or capacity. Uses everyday language related to time. Beginning to use everyday language related to money. Orders and sequences familiar events. Measures short periods of time in simple ways.	Ask "silly" questions, e.g., show a tiny box and ask if there is a bicycle in it. Be a robot and ask children to give you instructions to get to somewhere. Let them have a turn at being the robot for you to instruct. Encourage children to use everyday words to describe position, e.g., when following pathways or playing with outdoor apparatus.

We think the most important thing to remember is that people learn how to measure by measuring. Sometimes we learn this by observing. We learn to bake and cook by watching someone we love and respect baking and cooking and by working alongside them. We learn to put up shelves or carry out more complex constructing work by working alongside someone and gradually taking on more responsibility. Activities in school can mirror this. It is worthwhile and enjoyable to build, measure, compare and explore using measures alongside the learners. Introduce them to the language of measure by using appropriate terms consistently, show your multilingual learners that you respect their family's background by learning key vocabulary in their home language. Learn alongside the children. Develop displays both as posters and as collections and label these. Use the mathematical vocabulary to develop reading and writing skills.

Another way that you can support your young learners in becoming mathematical is by developing a more mathematical view of the world yourself. Go for a walk around the school. Notice all the different shapes that you can see. You will come across shapes that you don't know the name off. Take a photograph or sketch the shape and find out what it is called when you get back to your computer.

Young children learning geometry

Tony has been a member of the Association of Teachers of Mathematics (ATM), in the UK since he first started teaching in 1981. One of the guiding principles of the ATM says:

The ability to operate mathematically is an aspect of human functioning that is as universal as language itself. Attention needs to be drawn to this fact constantly. Any possibility of intimidating with mathematical expertise is to be avoided.

The students in the large secondary school in the North of England where he started his career had spent a large part of their time being intimidated by mathematics to the extent that they decided they were not capable of thinking mathematically. Indeed, if they were successful in any of the lessons, which they were, they would decide that what Tony was introducing them to could not be mathematics. Mathematics, for them, was defined as that stuff they couldn't do.

Tony recently commissioned a very successful sculptor and graphic artist to create a graphic for the mathematics journal he edits. He asked him to read through the articles and create an image that summarised his response to the articles. When they met the artist said he felt unable to this as he couldn't understand all the articles. He said, "I loved the articles about geometry and using pattern but I couldn't understand the one about prime numbers, so I stopped thinking." For him, it seemed, the articles exploring shape and space were not "proper" mathematics in the way that prime numbers are. He was intimidated by mathematics or by what he perceived as "real" mathematics.

We want to try to adjust your metaphor for the world of mathematics. If you currently see mathematics as a staircase or a ladder and the aim in learning mathematics is to ascend this ladder as it disappears into the clouds, erase this image. Instead, imagine a map of a large country. This country contains many different areas to explore, and you can discover many routes around the country. One of the main states in this country is called "shape and space" and this is just as important a part of the country as the county called "number." People who live here and who spend time here are just as much members of the mathematical community as people who spend all their time in "number."

What might you find in your exploration of this fascinating place? The following table will start to fill in the map:

Age range	What a child is learning	What adults could do
Birth–11 months	Babies' early awareness of shape and space grows from their sensory awareness and opportunities to observe objects and their movements and to play and explore.	Provide a range of objects of various textures and shapes in treasure baskets to excite and encourage babies' interests.
8–20 months	Recognises shapes in meaningful contexts.	Encourage babies' explorations of the characteristics of objects. Talk about what objects are like and how objects, such as a sponge, can change their shape by being squeezed or stretched.

Age range	What a child is learning	What adults could do
16–26 months	Attempts, sometimes successfully, to fit shapes into spaces on inset boards or jigsaw puzzles. Uses blocks to create their own simple structures and arrangements.	Use "tidy up time" to promote logic and reasoning about where things fit in or are kept. Help children use their bodies to explore shape, through touching, seeing and feeling shape in art, music and dance. Use "shape-sorters" and similar common toys to explore shapes.
22–36 months	Notices simple shapes and patterns in pictures. Begins to categorise objects according to properties such as shape. Begins to use the language of shape.	Talk about and help children to recognise patterns. Draw children's attention to the patterns e.g., square/oblong/square that emerge as you fold or unfold a tablecloth or napkin.
30–50 months	Shows an interest in shape and space by playing with shapes or making arrangements with objects. Shows awareness of similarities of shapes in the environment. Shows interest in shape by sustained construction activity or by talking about shapes or arrangements. Shows interest in shapes in the environment. Uses shapes appropriately for tasks. Begins to talk about the shapes of everyday objects, e.g., "round" and "tall."	Demonstrate the language for shape, position and measures in discussions. Find out and use equivalent terms for these in home languages. Encourage children to talk about the shapes they see and use and how they are arranged and used in constructions. Value children's shape patterns, e.g., helping to display them or taking photographs of them.
40–60+ months	Begins to use mathematical names for "solid" 3D shapes and "flat" 2D shapes and mathematical terms to describe shapes. Selects a particular named shape. Uses familiar objects and common shapes to create and recreate patterns and build models.	Play peek-a-boo, revealing shapes a little at a time and at different angles, asking children to say what they think the shape is, what else it could be or what it could not be. Introduce children to the use of mathematical names for "solid" 3D shapes and "flat" 2D shapes and the mathematical terms to describe shapes.

These activities lead us to the content descriptions, "Sort, describe and name familiar two-dimensional shapes and three-dimensional objects in the environment." This appears to make an Early Years educator's job very straightforward – provide children with the opportunity to observe and create patterns and to describe the world they observe using the language of geometry.

The centre of the experience here is noticing what is the same and what is different about objects. As we describe shapes and objects that we see, hold and move we begin to use mathematical vocabulary. As we classify shapes and objects, we use mathematical vocabulary to justify our different classifications. It is important that Early Years educators become confident in using mathematical vocabulary about shapes. Introduce the mathematical names for shapes as soon as children start to notice and describe them. It is also important to model the correct use of shape names when talking with children. For example, not calling a square a diamond if it tilted 45 degrees. Students will need some assistance to see and understand why this shape remains a square and thus should retain the same name. If you don't know the name of a shape, don't worry, look it up and be as excited as your young learner when you discover what it is called.

We hope that this section gives you the confidence to plan activities involving noticing, sorting, classifying and describing shapes and that you are beginning to see mathematics as much more than simply counting.

Young children learning algebra

Overt discussion of the exploration of algebra, both in terms of numbers and shapes, does not appear in the content descriptions apart from general statements about exploring and noticing number patterns. However, we think algebra is worth separating from the previous areas so that we can focus on the specific algebraic skills that children are developing as young learners. You will notice that we have deliberately not used the phrase "pre-algebra." Our view is that, if young learners are coming to understand ideas that can be described as algebra, then they are doing algebra.

TAKING IT FURTHER: **FROM THE CLASSROOM**

In the article, "Algebra in the Early Years? Yes," Jennifer Taylor-Cox argues for the inclusion of algebra in the Early Years curriculum. She describes how, in the United States, algebra acts as a gatekeeper for access to universities and higher education. In Australia, many learners see algebra as symbolising all that is challenging about mathematics. If you understand algebra you belong to that special group of people or are "good at mathematics." So, what does algebra in the Early Years look like?

The moment we start to notice patterns and either describe them or continue them we are thinking algebraically. This might be patterns of shapes or staircases built using Cuisenaire rods. If we notice a general rule that we can use to describe a pattern, we are using algebra to describe the world.

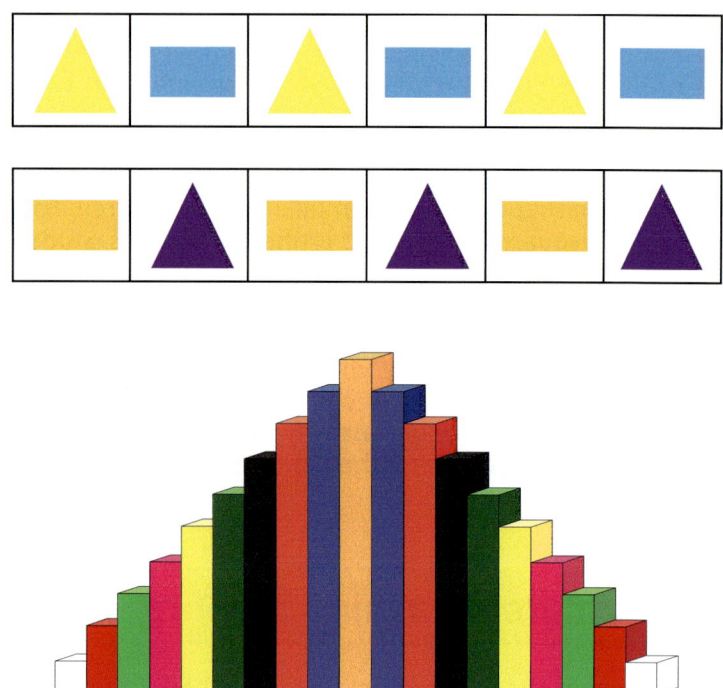

When you use balance scales to compare masses of objects, you are exploring algebra. A young learner may discover that three rolls of tape balance two lemons or that four toy cars balance two trucks.

Encourage your young learners to record these relationships in any way they wish. Again, we see that mathematics in general and, in this case, algebra in particular is something that happens all the time, all around us. Our main task is to notice and share the mathematics that is all around us.

Creating a mathematically rich environment

We think one of the most important things an Early Years educator can do is to develop an environment within which young learners feel secure and able to express their curiosity. Think, for a moment, about a time when you really felt you were learning. When you were exhilarated by the progress you were making, when you discovered something new and exciting, when something you thought you were certain about was challenged and you had to adjust the way you thought about the room. Put yourself back in that place with those people and try to bring that sense of inspiration to the front of your mind.

Where were you? Were you inside or outside? What was the weather like – what about the decoration in the room? What equipment were you using? What resources were available? What "things" could you see around you? And finally, who were you with? Who was inspiring you? Supporting you? Challenging you? Questioning you?

These are all questions about the learning environment, and you have just re-imagined what makes a rich learning environment for you. This chapter explores what it is that makes not just a rich learning environment but a mathematically rich learning environment. There is much in common with the learning environment you have just imagined, but there are also things that are specific to create the conditions for learning mathematics.

A mathematically rich environment consists of three inseparable parts: people, spaces and objects. We will explore people and spaces in detail later, discussing the particular role they play in creating a mathematically rich environment. Then we will return to the Early Learning goals to develop the ideas contained there about developing a mathematically rich environment linked to the specific areas of mathematics. But first we need to agree that the environment matters.

Creating a mathematically rich environment

Belonging, being and becoming suggests that early learning experiences are best delivered through partnerships between educators and families, recognising that all children's first educators are their families. It lists key facets of successful partnerships as:

- valuing each other's knowledge of the child
- valuing each other's contributions
- trust
- communicating freely and respectfully
- sharing insights and perspectives about the child
- engaging in shared decision making about the education of the child.

We can probably agree that such partnerships are valuable through-out a students' educational career, but perhaps they are even more important at the earliest stages. What sort of a mathematically rich environment might educators try to develop? *Being, belonging and becoming* suggest that the environment should be a welcoming space that reflects the partnerships described earlier and caters to the diversity of learning styles and children present in the environment.

There is particular emphasis on outdoor learning, which, they suggest, is a particular feature of Australian education. They describe these spaces as occurring in

> *natural environments include plants, trees, edible gardens, sand, rocks, mud, water and other elements from nature. These spaces invite open-ended interactions, spontaneity, risk-taking, exploration, discovery and connection with nature. They foster an appreciation of the natural environment, develop environmental awareness and provide a platform for ongoing environmental education.*

As we have said before there is very little mathematics specific advice in the Early Years guidelines. So, what skills and understandings might we expect adults, in the settings described earlier, to display if they are to add to the richness of the environment? *Children thinking mathematically: Essential knowledge for Early Years' Practitioners*, a report on research commissioned by the English government, lists four important characteristics. The document suggests that mathematically rich environments are supported by educators that:

- share positive beliefs about young children learning mathematics
- are aware of the mathematics that arises through children's self-initiated play
- have high expectations of young children's mathematical understanding
- understand babies' and young children's mathematical development, learning from reflecting on observations and through discussions with their team and use this knowledge to "tune into" the mathematics that children explore within their play.

We suggest you use this list as an aide-memoire for yourself or for your colleagues working in the Early Years, asking them to reflect honestly. If we take the time to observe young children engaged in mathematical activity, we cannot fail to become aware of the mathematics that naturally arises through self-initiated play. It really is only a matter of looking for it. It is though noticing the mathematics within children's play that you will be able to share your positive beliefs with the children. You will have to ration yourself to the number of times you say, "What an amazing piece of mathematical thinking!"

The final two bullet points do demand a bit of work on our behalf. We think all Early Years' educators are brilliant at planning and supporting activities across most of the curriculum. However, we are not convinced that all practitioners are confident enough in their own mathematical understanding to challenge young learners thinking and or to understand mathematical development. Hopefully, if you have already the chapters preceding this one, you are feeling more confident. I am sure you will have high expectations of all the young children you work with. This type of adult is described in *Children thinking mathematically* as someone who:

> Value[s] children's ideas and support[s] children's play and mathematical explorations through collaborative dialogue help to 'scaffold' children's thinking. Practitioners can help children go beyond what they already understand and can do. Thinking, making meanings and understanding are significant aspects of mathematics.

The people in a mathematically rich environment also see children's learning as a partnership between the setting and the home environment. We should note that any mathematically rich environment would include educators talking with parents about their children's developing understanding of mathematics. Our ideal environment would also include educators, parents and children all actively exploring mathematics together. This is supported by the description of partnerships in *Being, Belonging and Becoming*.

Assessment in the Early Years

Effective assessment in the Early Years is an analysis and review of children's learning in order to make informed decisions about what you will plan to meet their next learning needs. This is the basis of assessment for learning: you are not assessing simply to find out what a young learner can do, you are observing their skills and understandings so that you can decide what you do next. There is a range of things you might look out for. Some educators choose to keep notes in a journalists' notebook during the day; you will see some educators using sticky notes and photographs so that they can reflect at the end of the day on children's individual progress and the next steps in learning they should provide. The following sections provide you areas to focus on when you are assessing your students' current understandings in order to plan for their next stage of development.

Numbers as labels for counting

As you sing number songs and rhymes with actions with the children, make a note of their responses. Observe how confident the children are with their actions. Do they initiate the actions or are they following their friends without understanding what the meanings of the actions are? When you count with children using displays, observe whether the children point at the numbers on the display and name them accurately and see what range of numbers they refer to?

You will be estimating small numbers of objects and can note down how accurate their estimates of numbers of things are. Similarly, when you use the language of comparison on a day-to-day basis you can note when children begin to use the language of first, second, third. You could also make a note of how the children refer to numbers they use in everyday life – house number, number of brothers and so on.

When carrying out counting activities with the children you can take note of their developing skills in this area. At what stage do they understand the one-to-one principle, the stable order principle; when do they realise that the last number in a count always gives the total? (See Chapter 4.)

Calculating

As you observe and engage the children in activities involving calculating you can look for how accurately children share things out between them. See if the children share accurately and make note of how they go about sharing.

Observe the children to see if they know whether or not the sharing has been fair by noticing that groups of objects have the same number or different numbers of objects in them.

You can pose questions whenever you have planned a "sorting" activity. See if the children work out how many more or less there are in some groups, and notice how the children respond if you combine two groups and ask them how many altogether. Alternatively, if you remove objects from a group, what forms of records are they developing to record your actions?

Measurement and geometry

At an early stage you will be noticing how the children respond to shape sorting games and the language they use to describe shapes. Listen to their vocabulary when describing measurement activities and the properties of shapes. You will want to listen out for language such as "bigger" and "smaller" or "heavier" and "lighter" or "in front" and "behind" as well as simple shape properties and names.

Make a note when children start noticing shapes in the environment or responding to you, pointing out shapes around the classroom. You can also observe the children working on practical activities such as modelling or wrapping presents, to see if they are estimating sensible amounts of materials. You can talk about where their lockers or tubs are in relation to each other's' e.g., my locker is next to Max's, it is also above Isabel's.

Do not see any of the aforementioned questions as "testing" the children. You can pose these questions, engage the children in the activities we have described and observe how they respond. As children become more familiar with the activities and as they hear their teachers and friends model more complex mathematical vocabulary, you will notice the children develop their understandings.

These are examples of activities which are rich in assessment opportunities. The first will help you explore children's understanding of counting:

- create a set of "lily pads," numbered 1–20, that the children can stand on
- ask the children to put them in order so that you can play a game.

It is very often mathematically worthwhile for the children to help you prepare for an activity. You can notice how the children order the lily pads:

- who takes control?
- who decides on the correct order?

Then ask the children to start on 1 and hop on the lily pads until they reach 15 or start on 12 and hop back to 2. For some children you could ask them to hop in 2s and others could count out loud the numbers they land on. You can also create number frogs; these are frogs with numbers on them that children have to match to the correct lily pad. Alternatively, children can make their own number frogs.

This activity allows you to assess children's developing understanding of shape, space and measurement. Provide the children with a large collection of newspapers, sticky tape, paper, modelling straws, pipe cleaners. The wider the range of resources, the better. Initially, talk about all the different sorts of houses the children can think of. Ask the children which houses are big, small and medium sized. Try to get children to explain how they are making the decisions about which category to place a house in. Give the group three different animals. Their task is to make a house for each of the animals. This assessment task relies on the teacher observing the process very carefully. It may be appropriate to use a digital camera to record the activities the children are involved in. Make sure you ask probing questions like:

How do you know that will be big enough?
How could you check the animal will fit?
How big is that house?
Which house is the biggest?

Summary

This chapter has outlined how problem solving and reasoning underpins numeracy learning in the Early Years. It has also emphasised how your teaching should engage the young learners in games, songs and active learning through which their understandings of number and shape and space will emerge. The importance of playful activity for mathematics learning is revisited on several occasions.

There are suggestions for the early and developing skills that you should focus on at this stage and a range of foci for assessment are offered. There is a clear emphasis on assessment for learning; that is your assessment is made to inform your plans for the next stages in learning for individual learners.

Reflections on this chapter

We hope that the chapter has shown how mathematics in an Early Years setting can appear as natural and everyday as all the other activities across the other areas of learning for the children. The curriculum and activities that you can plan and provide should engage and challenge the children – and hopefully you will enjoy engaging with the children. Children are developing their sense of identity as they work with you – if part of that identity can be that they see

themselves as successful in mathematics and able to tackle mathematical problems, you will have succeeded as their first mathematics
teacher.

Going further: further reading

How to develop confident mathematicians in the Early Years

TONY COTTON

This chapter has drawn heavily on this book, which is specifically
aimed at Early Years practitioners. The book shows how everyday
experiences can be used to develop young learners mathematical
thinking and is aimed at parents as well as practitioners. The second
half of the book contains 60 activities that can be used on an Early
Years setting.

Cotton, T. (2019) *How to Develop Confident Mathematicians in the Early
Years*. London: Routledge.

Belonging, being and becoming: the Early Years framework for Australia

AUSTRALIAN GOVERNMENT DEPARTMENT OF EDUCATION

This is the first such framework and aims to extend and enrich children's learning in the Early Years of their development up to the transition to school.

Available at www.dese.gov.au/national-quality-framework-early-
childhood-education-and-care/resources/belonging-being-becoming-
early-years-learning-framework-australia

Mathematical problem solving in the Early Years

SUE GIFFORD

Sue Gifford suggests that young learners are natural problem posers
and problem solvers. The article suggests way that practitioners can
support children in developing their problem solving skills.

Gifford, S. *Mathematical Problem Solving in the Early Years: Developing
Opportunities, Strategies and Confidence*. Available at https://nrich.maths.
org/12166 Accessed 2 March 2018.

Early mathematical experiences

PAUL SWAN

This book is designed for teachers working with four- to six-year-old children. It includes a variety of "play-based" experiences that lay the foundation for learning mathematics. Some examples of experiences are using jigsaws, sand and water play, construction and cooking.

Available at https://drpaulswan.com.au/shop/early-mathematical-experiences/

Teaching and learning early number

IAN THOMPSON

The most complete guide for Early Years practitioners focusing on the learning and teaching of number. Whilst this is focused on the research, it is accessible and an important guide to the theory of the beginnings of learning to count and to calculate.

Thompson, I. (2008) *Teaching and Learning Early Number*. Oxford: Oxford University Press.

CHAPTER 12
ISSUES OF INCLUSION

Introduction

In 2018, whilst Tony was working on the UK edition of this book, he was supporting the son of a friend to prepare for his mathematics examination at the end of his secondary schooling. The son had been assessed as having Asperger's syndrome, although this term has been subsumed into the overall term Autism recently, and this has led to him missing large parts of his schooling. He has also been placed in special units for substantial amounts of his time at school. Unsurprisingly, his mock examination results suggested that he was unlikely to gain the grade that is necessary to progress into higher education as many colleges and universities now require a grade "C" or equivalent for a place on any course.

After a short intensive period of tutoring, John achieved a grade "D" – just below the required grade "C" but a great achievement as he gained "just about zero" in his revision assessments at school. However, the College of Music who had offered him a place on a music production course withdrew the offer, which was a huge disappointment. Tony and John worked together for another year, and at the end of this year he successfully gained the grade C he needed. Unsurprisingly, however, John has chosen not to return to an education system that has not been supportive to him and is now developing a career in sound engineering through finding employment of his own.

The most recent research, from the organisation Ambitious about autism, suggests that only 15% of adults with autism are in full time, paid employment. Of the people polled, 79% said that they would like to be in full time paid employment but had been unsuccessful in finding jobs. This episode reminds us of the huge influence that learning and teaching mathematics has and the particular struggles that young people with special learning needs face.

DOI: 10.4324/9781003315155-12

Understanding and becoming successful in mathematics can have a huge influence on individual life opportunities. None of you would have been able to gain places on teacher education courses without success in Year 12 mathematics.

In 2018, the organisation Amaze, a group that supports people with autism in Victoria, commissioned two research reports, *Community attitudes and behaviours towards autism* and *Experiences of Autistic people and their families*. These studies were conducted by the social research centre at Australian National University and the Centre for Health and Social Research at the Australian Catholic University. This research showed that only 35% of autistic students make it as far as Year 10 compared with 17% of the general population. They also found that many autistic students change schools multiple times (24% in primary school and 44% in secondary schools). This means that many of these students' learning needs are not met.

Starting point: further reading

In 1997 Mike Askew, Margaret Brown, Dylan William and colleagues from King's College in London explored the links between teachers' practices, beliefs and knowledge and student learning outcomes in mathematics. They interviewed and observed 90 teachers and 2,000 students. The full results are available in the book *Effective Teachers of Numeracy: Report of a study carried out for the Teacher Training Agency* (King's College).

In this book the authors identified three sets of beliefs that they suggested were important when understanding the impact of teacher beliefs on effective teaching of numeracy. These were:

1 **Connectionist:** a connectionist teacher values students' methods and teaching strategies with an emphasis on establishing connections within mathematics. This means that learners are able to see the links between the different areas of mathematics they are engaged with and can see the "big picture" rather than view mathematics as a set of separate skills to be learnt in isolation from each other

2 **Transmission:** a teacher with these beliefs sees mathematics as a collection of separate routines and procedures to be taught to students

3 **Discovery:** these teachers see themselves as facilitators of learning and see mathematics as an area to be discovered by students.

You will recognise these terms from the audit questionnaire you completed in Chapter 2. The research showed that teachers with a strong connectionist view were most effective in terms of students making progress in their learning of mathematics. A key belief for connectionist teachers is that "Most students are able to become numerate." This means that they believe that all students are able to move forward in their mathematics learning. The challenge for teachers is to find the most effective ways for students to learn mathematics. This chapter explores the ways in which you can adapt and develop your practice to include all children in your class, whatever their learning needs are.

Chapter 3 opened with this number square:

1	3	5	7	. . .
2	6	10	14	. . .
4	12	20	28	. . .
8	24	40	56	. . .
. . .	. . .	. . .	. . . . . .	

We posed the questions, "What patterns can you see?" and "Will 1,000 appear?" Tony has used this activity with learners aged 8–80, with primary school students, secondary school students and degree-level mathematicians. Some groups of students focus on exploring patterns. The simplest patterns here are odd and even numbers. Others have found series of numbers that increase by 2, by 4, by 8 and so on. Another group noticed that the numbers in the first column are powers of 2 ($2^2 = 4$, $2^3 = 8$ and so on). Older students used algebra to explore the patterns in the square, and academic mathematicians came up with a proof that showed all numbers would appear once and only once.

As a teacher exploring ways of teaching for inclusion through mathematics, a key skill you will need to develop is the ability to plan activities that are accessible to the range of students in your class. However, as well as offering accessibility, these activities also need to offer challenge. Many students with specific learning needs describe lessons spent colouring in hundreds chart or drawing pictures as teachers did not expect them to rise to any challenges. Needless to say, these students' memories of mathematics lessons are of lessons filled with boredom.

Another activity you were introduced to in Chapter 10 might engage a wide range of students. If you remember, students were asked to try to think about the relationship between the distance around their head and their height and come up with some conjectures. If students work in all attainment groups for this activity, they can bring their different strengths to bear on the investigation. In this activity, the students measure each other's height and the circumference of their heads to create a scatter graph using initials rather than dots so that each student can see themselves in the data. A scatter graph is a graph of plotted points that shows the relationship between two sets of data. For this activity, distance around head was measured on one axis and height on the other.

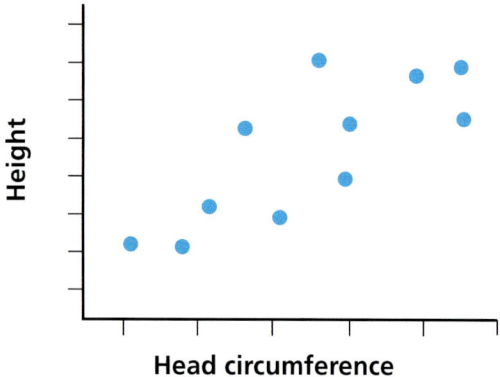

Head circumference

Different groups will interpret the graph in different ways. Some find the "line of best fit." The line of best fit is the line that can be drawn on a scatter graph, which is the best approximate to all the points on the graph. If you have drawn it accurately, the total distance of all the points from the line should be the same for points on each side of the line. Some students used calculators and individual measurements to see if there was a pattern; some looked at boys and girls in the data to see if there were generalisations that could be made. But the whole class was engaged in exploring the data. At the end of this activity in a classroom Tony was working in, the teacher turned to him and said, "That was interesting. The weaker mathematicians were the most successful." In this class, the teacher usually grouped her children by their prior achievement in mathematics tests. These tests were usually number based. The teacher had noticed that for this activity prior achievement in a number test did not predict how well an individual would engage with ideas if data handling and problem solving.

In an inclusive classroom all students will be achieving to the best of their ability, and different students may well excel in different areas of mathematics rather than individuals attaching labels of "good at mathematics" to themselves. A key lesson here is how important it is to vary the groupings we operate with in schools. When we visit schools, many still operate with fixed groups for mathematics lessons. If we never vary the groups that students learn mathematics in, we limit the range of their peers that they can learn from, and we also limit the possibility for children to excel in different areas of mathematics as they quickly come to understand they are in the "bottom group" for mathematics and put limits on their own understanding.

This became clear for Tony one day when he was sitting with a group in a Year 2 classroom. One boy in the group said to him, "Can you do all of your times-tables?" Tony told him he could, so the boy said, "You need to go and sit with the blue group. They are the cleverest and we won't know our times-tables until next year."

TAKING IT FURTHER: **FROM THE RESEARCH**

Professor Jo Boaler, whom you have met in previous chapters, has researched the impact of forming groups based on previous performance (e.g., scores in a test taken at the start of the year). She suggests that this practice means that many students spend much of their time in school being given low-level and uninteresting activities. Nearly all of these students are capable of achieving more if they are given appropriate learning activities.

In interviews she carried out she was moved by the pleas of students in "low-ability" groups who said that they just wanted to learn but were constantly given activities that they did not feel stretched them. She notes that some people believe the practice is right because it keeps the high achievers away from low achievers. However, her research shows that high achievers do not do any better in "high-ability" groups than in mixed ability groups, and for some students being in the "top group" is a source of considerable anxiety. She also shows us that comparisons of test performance in different countries always show that countries that do the least grouping according to performance and that leave it until the latest time to do so have the highest performance. The reasons for this are obvious: once students are told that they are low achieving and are given low-level work, their learning diminishes.

This conclusion is supported by the Education Endowment Foundation. The teachers' toolkit, which is available on their website (https://educationendowmentfoundation.org.uk/evidence-summaries/teaching-learning-toolkit/) suggests that setting or streaming, in other words grouping by performance, has a negative impact on attainment. They have found that, "on average, pupils experiencing setting or streaming make slightly less progress than pupils taught in mixed attainment classes."

Jo Boaler's conclusion is that a mixed ability, multi-dimensional approach means that all students can succeed. Jo has launched the organisation *youcubed*. You can find this at the website www.youcubed.org. Their mission states that:

> *All students can learn mathematics to high levels, and teaching that is based upon this principle dramatically increases students' mathematics achievement. The need to make research widely available is particularly pressing now as new science on the brain and learning is giving important insights into mathematics learning.*

If you follow this belief, you will be working to close the achievement gap. The rest of this chapter focuses on ways in which teachers can meet the specific needs of particular groups of students who may currently be underachieving in mathematics.

Children with special educational needs

It is important to remember that there will not necessarily be a correlation between students you teach who are identified as having a special educational need and their ability at mathematics. For example, a student in your class who is dyslexic may be a high achiever in mathematics or may find it more of a struggle. There is often no direct correlation between a special educational need and attainment in mathematics. However, there are steps you can take that ensure you are giving all children the best possible chance to achieve their potential. It is vital that you become aware of the impact of an individual student's specific need and use this to support your planning.

The main implications for your planning are ensuring that activities are accessible for all students so they can find a way to get into the activity, whilst ensuring the activities offer sufficient challenge for all students. This might mean that you have to adapt resources that have previously been used to teach a particular idea. For example, children who are learning about adding money to one dollar may be given the following questions:

Copy and complete these number sentences

20c + _ _ = $1	90c + _ _ = $1
40c + _ _ = $1	50c + _ _ = $1
70c + _ _ = $1	10c + _ _ = 1$1

This closed activity is accessible to students who already have the necessary skills, but other students will find it hard to make a start on the activity. Similarly, students who understand the task will find it straightforward and complete the activity without being challenged.

An alternative to this task is for the teacher to ask, "How many different ways can you make $1?" ensuring learners have access to 5c, 10c, 20c and 50c coins to support them if they choose to use manipulatives to solve it. For students who are excelling at this task and finding multiple combinations, they could be asked to figure out how many answers there are to this problem. Try this activity for yourself and you will see that it offers accessibility and flexibility. Tony reports on using a similar activity at parent workshops to show how more open activities can be both accessible and challenging. It also illustrates that such activities can offer more "practice" as this is often important to parents.

If you plan the structure of your lessons carefully you can maximise the possibility of all your students progressing in their learning. For example, in the sections of the lesson that focus on discussion led by the teacher, it will help to plan questions specifically for students who are being challenged by the content of the lesson. By careful targeting of questions rather than asking for "hands up," you can draw all students into the activity. This also allows you to focus on and develop key vocabulary. Another good way to involve all students in mathematical discussions is by encouraging conversations between groups of students during the lesson. This process of collaboration through discussion supports all students in the room, from those who are finding the task at hand challenging to those who are progressing confidently.

Careful thought about the resources you can use will also support your students with specific learning needs. Wall displays that have been co-constructed with the students will support them in developing appropriate vocabulary. The use of number lines, hundreds charts and other visual tools will also support children to develop mental images of the ways in which the number system works. There will be students you will meet in your career who have been identified as having specific learning needs. There are suggestions later as to how you may support these particular groups of students. Don't forget, however, that most of these ideas will improve the learning experience for all your students.

Gender issues

A few years ago, Jess asked the Year 3 and 4 students she was teaching at the time to visualise and draw a mathematician, including a sentence explaining who and what they had drawn. Who and what do you visualise when you think of a mathematician? This also begs the question, what do you think makes someone a mathematician? Perhaps the results of this exercise won't surprise you. In terms of gender, all of the boys in the class depicted a male mathematician. About half of the female students also illustrated a male mathematician. It's also interesting to note that of the girls who drew females, almost half of those drawings were of teachers.

You might like to try this exercise with your own students. We suggest you unpack the results together and use it as an opportunity to discuss gender stereotypes and biases, as well as real-life STEM careers.

At first it might appear odd to have a section dedicated to the inclusion of girls when females make up half of the world's population. And yet although this is true, gender remains a significant factor impacting on achievement in mathematics, attitude about mathematics and participation in non-compulsory maths subjects, diminishing later career options and subsequent economic opportunities. Key findings from the latest Australian STEM Equity Monitor, a national data report, support this last statement. Though there are improvements, the 2019 data revealed that women still make up less than a quarter of students studying STEM at a university level. This fact remains a significant contributor to the gender pay gap as a large proportion of STEM professionals earn an income at the top of the pay scale. The Australian Mathematical Sciences Institute (AMSI) has many useful resources on their website dedicated to careers in STEM with an emphasis on female representation.

TAKING IT FURTHER: **FROM THE RESEARCH**

In the Mathematics Teaching Toolkit (MTT) published by the Victorian Department of Education, Helen Forgasz and Gilah Leder report on Issues in the Teaching of Mathematics: Gender and Mathematics. They use evidence to highlight two main areas to consider: performance and participation.

In terms of performance, over the past decade boys have, on average, slightly outperformed girls on the mathematics section of the NAPLAN test. This is the case for each year the test has been administered and at each grade level. And although there was a slightly higher proportion of girls than boys meeting the National Minimum Standard in each grade level in 2019, a higher proportion of boys scored in the highest band in each grade level that year. This is a finding consistent with previous years on the same test, as well as outcomes reported in other countries.

	Mean NAPLAN score		% at or above NMS1		% at or above highest band	
Year	M	F	M	F	M	F
3	421.5	403.5	95.1	96.0	19.1	14.0
5	501.9	489.6	94.8	96.0	12.4	7.5
7	555.7	549.8	93.8	94.9	16.5	12.1
9	597.0	587.0	95.5	96.5	9.5	5.8

Regarding participation, Forgasz and Leder look at the enrolment in the three elective VCE (the senior secondary school qualification in the state of Victoria) mathematics subjects. For each year (2001–2019) and for each of the subjects, a notably higher proportion of boys than girls were enrolled. Again, similar findings are reported in many other countries.

Students' own beliefs and attitudes are believed to affect their performance and desire to pursue further mathematics education. Some of these beliefs have been outlined in Forgasz and Leder's Mathematics Teaching Toolkit:

- more boys than girls say they like doing mathematics (though for both groups, liking of mathematics decreases as students move into higher grades)
- more boys than girls are confident they can do well in mathematics
- more boys than girls indicate that their parents and teachers expect them to do well in mathematics, though in some recent surveys many girls also say that they believe girls are as good as boys at mathematics
- when shown a mathematics question, more boys than girls state that they can solve it correctly
- more boys than girls expect to use mathematics in their work
- more boys than girls volunteer to answer or ask a question in class
- more girls than boys say they like to work with others when doing mathematics.

Another contributing factor points to a disparity in experienced maths-anxiety. Girls not only experience higher levels of maths-anxiety than their male classmates, but it also starts younger for girls and evidence suggests that it affects their performance to a greater extent as well. Maths-anxiety (for any student) often leads to avoidance, lack of confidence and general disengagement from the subject.

We must also consider how we as teachers may unintentionally play a role. Studies of interactions in mathematics classes have revealed that teachers, on average, spend more time with boys than with girls, interact more often with boys than with girls, more regularly ask boys more challenging questions and girls simpler questions and offer boys longer think times than girls. What messages are we inadvertently sending to our students when we do this? As the Numeracy Learning Specialist at her school, Jess sometimes carries out classroom observations specifically monitoring this particular aspect of a lesson, often leading to eye-opening results. We encourage you to become more attentive of your own unconscious biases in the classroom.

To close the gender gap in mathematics achievement and attitude, you need a whole-school approach that includes families and the community. Messages including that it's okay not to understand something straight away and rewarding hard work are key to building confidence; fostering a growth mindset will help minimise maths-anxiety in all students. Eliminating biases by becoming more aware of our own will begin to challenge engrained stereotypes. By emphasising real-life STEM applications and careers you're exposing your students to authentic role models and opening up exciting pathways. All of these suggested adaptions to your teaching practice will not only improve girls' performance and engagement in mathematics, it will provide opportunity for economic growth and overall social justice.

Children learning English as an additional language (EAL)

Considering how to make our classrooms good places for learning mathematics for students who are learning English as an additional language is a good example of how a special educational need should be seen as a benefit to the classroom rather than being a problem for teachers and students. Our understanding of language allows us to both communicate and think about our understandings of the world. Whenever we work in multilingual settings, we learn new ways of looking at the world that are not available to monolingual teachers and students through English alone. Those of you who are multilingual will understand this. You will be able to think of examples of words that describe ideas in a language other than English that do not directly translate. Those of you who are monolingual should talk to multilingual colleagues or friends who will quickly offer you examples, and you will begin to understand how much richer the world is when we see it through a multilingual lens.

The importance of seeing multilingualism as a strength was brought home to Tony when he ran a series of parent workshops at a primary school in Leeds, England. The sessions were aimed at the parents of seven- and eight-year-old students who were beginning to struggle a little in mathematics. It was hoped that this early intervention would stop the achievement gap from opening up.

The first session took as its theme number recognition. Tony wanted to begin the session by focusing on the diversity of experience in the room. He was aware that the parents' experience of schooling would not always have been positive and that for the majority for the parents their experience was in schools outside England. So, he began by sharing the languages that there were in the room. All the participants, including Tony, learnt to count in Czech, Albanian and Mandarin and the parents completed a bilingual number track in their first language. This allowed the parents to be in a position of knowing. It also gave Tony an idea of the parents' personal starting points both mathematically and linguistically. He then promised to learn the numbers 1–20 in one of the languages for the future sessions. The parents took a pride in teaching their home languages and, as Tony was placed in the position of learner, communication was opened up in an open, accessible, multilingual setting.

TAKING IT FURTHER: **FROM THE RESEARCH**

In the *Language, Mathematics and English language learners*, Misty Adoniou form the University of Canberra and Yi Qing from the ACT Education and Training Directorate explore the challenges in learning mathematics for students who are also learning English. Their research provides an overview of the linguistic features of mathematical language and the particular language challenges in learning the language of mathematics. Most importantly they suggest some strategies that can support teachers in teaching English and the specific language of mathematics simultaneously.

They suggest that 15% of learners are disadvantaged in mathematics lessons because of language issues rather than issues with the mathematics itself. They also find that many teachers have not been offered training or support to help them better meet the needs of those students who are learners of English. The problems for students include decoding mathematics problems at sentence level – that is they can see the numbers within a problem but cannot work out what they are supposed to do with these numbers. This resonates with the reported experiences of colleagues of Tony's in the UK who found that their students with English as an additional language achieved more highly on assessments that were devoid of context and simply presented them with calculations.

Strategies that were suggested by the research included reading through problems together, identifying the key words and phrases critical to the mathematics. They suggest teaching language by analysing the text and comparing similar words such as "investigate" and "calculate." Revisit and practice new vocabulary whenever it is introduced and to take care to only introduce new vocabulary on the context of a mathematical activity that uses the vocabulary. Perhaps most importantly, they suggest that students work in pairs of in groups on activities so they can offer each other linguistic support.

It has also been shown that most learners with EAL develop a functional level of English very quickly. However, this does not mean that they will not need continued support to develop the proficiency in English, which is necessary for success with more complex learning that will come in later years. The continued use of active learning and language-rich activities such as matching activities, the use of manipulatives and discussion around mathematical language is important. In fact, this can be seen as important for all learners.

Perhaps most importantly learning a language is also about learning about cultures that may differ from one's own. This is another example of why a multilingual classroom is a blessing rather than a curse. It offers all learners access to a wide range of cultures. It offers multiple viewpoints and multiple ways of learning mathematics.

Children with social, emotional and mental health needs

Previous UK editions have used the terminology Emotional and Behavioural Difficulty (EBD), which was described as a disorder. This meant that to the behaviour or the emotional response of some children is so different from the rest of your class that it impacts on the child's learning. This may take the form of disruptive, antisocial or aggressive behaviour; poor peer relationships; or hyperactivity, attention and concentration problems. It seems more appropriate to see this as a learning need rather than a disorder. If we see something as a learning need it means that we are able to do something about it. If we pathologise the behaviour as arising from a "disorder" there is little we can do to impact the learning experience of this child.

You may find that unstructured, open-ended tasks can exacerbate the anxieties of some students. If children are uncertain how to start an activity or are unclear what your expectations of success are, they may avoid starting the activity in case they make a mistake or misunderstand what is expected of them. It may help to provide a clear structure and set time-specific targets. Try to be consistent in your routines, in particular the cycle of launching a task, providing time to explore and summarising. This is very important for some of the students who need to be able to see the same routines being followed. Giving students responsibility and involving them in demonstrations by bringing them to the front can also support some students to develop positive behaviour patterns as they take responsibility for their own learning and the learning of their peers. Aim to avoid interrogational questioning as this may lead some students to feel humiliated. This is a common cause of inappropriate behaviour. A useful strategy here is the use of turn and talk routines. Students can try out responses with their partner before offering their ideas to you and to the rest of the class. You can also listen in and encourage all students to contribute to a whole class discussion at the end of the turn and talk activity.

Children with autism

The group the Autism Association of Western Australia, on its website www.autism.otg.au, describe autism as characterised by differences in social skills, communication and behaviour. They go on to say:

> This means that people with Autism experience differences in the way they communicate and interact socially, and their behaviour may be repetitive or highly focussed

(the term "restricted, repetitive patterns of behaviour" is often used to describe this). People with Autism also tend to experience differences with their senses that can affect the way they feel about and respond to their surroundings. Autism is not a disease or illness.

The three main areas of challenge for all people with autism share are:

- **Social communication**: some people with autism have difficulties interpreting verbal and non-verbal language such as gestures or changes in tone of voice. You may notice students with autism taking everything you say literally. It is very helpful to students with autism (and many others) if you speak in a clear, consistent way avoiding sarcasm. It is also important that you give students time to talk about their own interests even though they may do this at length.

- **Social interaction**: some students with autism may appear insensitive to others or need to find a quiet space when they are overwhelmed by other people. Some may prefer to work on their own or find it hard to from friendships. None of this should be seen as inappropriate behaviour. It is much better to provided quiet spaces and allow students to take time out rather than put them in situations that may cause anxiety and lead to conflict.

- **Repetitive patterns of behaviours**: people with autism describe the world as unpredictable and confusing. They may well devise repetitive routines and behaviours as a way of controlling this confusion. It is important that once you have set rules and routines you try to stick to them as much as possible. Your unpredictable behaviour will be particularly confusing for a learner with autism.

It can be hard to create awareness of autism as people with the condition do not "look" disabled: parents of children with autism often say that other people simply think their child is naughty, while adults find that they are misunderstood. You will notice that we have avoided attributing certain behaviours to all students with autism. As with "neuro-typical" students, all students with autism are different. It is worth remembering the saying, "If you've met one person with autism, you've met one person with autism!" As a person with autism himself, Tony reminds us not to attribute stereotypical characteristics to all people with autism. However, the following strategies may prove helpful.

You need to be careful when setting up group or pair work as this can be problematic for some children on the autism spectrum. It is important not to avoid group and pair work, but you need to make sure that routines are followed consistently, in particular routines for moving into groups or pairs. For example, always sitting in the same pair on the carpet will help. Classroom transitions are another point of the lesson that can cause difficulties for children with autism. Plan for these transitions carefully, be very clear about how the transition will be managed and keep them calm and orderly. If you just send groups back to their tables without instructions this will prove difficult for children with autism. If they become overwhelmed by noise and movement they should be encouraged to go to a quiet, calm place for a while.

Using familiar equipment also supports these children in their learning as novelty has little appeal. Extensive use of metaphor can also cause difficulty as children with autism are likely to interpret what you say literally. Making time to observe the children to check their understanding is important. Using picture sequences to break down activities into manageable tasks can also support progression. Finally, allowing a student to "fiddle" with their fingers or small objects can help with concentration and focus.

Students with autism may find changes in classroom organisation confusing, so it is important that you signal any changes in routine or setting well ahead of time and explain clearly why you are changing things. Some students with autism are sensitive to bright lights or to sudden noises so think carefully about how you light spaces and encourage all students to avoid making sudden noises.

Films and stories in the media seem to suggest that many people with autism can calculate huge numbers in their heads and have photographic memories. Unfortunately, such stories give the false impression that all children with autism have these special skills. Most people on the spectrum do not have any special talents for mathematics. However, it can be true that children with autism will exhibit keen interests in particular subjects and so may know a great deal about very specific areas. The following suggestions for children with dyspraxia will also prove helpful for children on the autism spectrum.

Children with Developmental Coordination Disorder (DCD or dyspraxia)

DCD is thought to affect up to 10% of the population. It is caused by the way the brain processes information to help us organise our movements and affects the way that we plan what we are going to do and how we do it. DCD can lead to difficulties in terms of perception, language and thought. DCD does not impact on how well students are able to learn mathematics, but it will be helpful to support learners with DCD to ensure they reach their potential.

As with students with autism you should avoid giving a series of instructions for a task as this may well confuse them. This will mean that you may have to differentiate, giving some groups more complex instructions whilst breaking the instructions down for others. Giving "thinking time" is useful, for example saying, "In 30 seconds I will ask you to tell me three number pairs that sum to 100." Using individual whiteboards to rehearse answers will also support these students.

You may also be able to adapt equipment or resources such as using pens and pencils with special grips so that they are easy to hold.

Children with dyslexia and dyscalculia

Dyslexia is a specific learning difficulty that mainly affects reading and spelling. Dyslexia is characterised by difficulties in processing word sounds and by weaknesses in short-term verbal memory, and its effects may be seen in spoken language as well as written language. The current evidence suggests that these difficulties arise from inefficiencies in language processing areas in the left hemisphere of the brain, which, in turn, appear to be linked to genetic differences. It is widely believed that Einstein was dyslexic; this does not mean, however, that all dyslexic students will excel at mathematics. As with all learners, some will love mathematics and with the appropriate support will excel. Others will succeed with you as their teacher but may choose other areas of the curriculum for their enthusiasms.

The Australian Dyslexia Association (www.dyslexiaassociation.org.au) offers support for learners with dyslexia but also focuses specifically on dyscalculia. They interpret it in the following way:

Dyscalculia is a lifelong condition that makes it hard for individuals to perform mathematics related tasks. It's not as well known or understood as dyslexia. But some experts believe it's just as common.

Experts don't yet know for sure if dyscalculia is more common in girls or In boys. But most agree it's unlikely that there's any significant difference.

Individuals with a dyscalculia profile will have trouble with many aspects of mathematics.

Dyscalculia can cause different types of mathematical difficulties that may vary from individual to individual. It also often looks different at different ages. Dyscalculia tends to become more apparent as children get older. Symptoms can appear as early as preschool.

They often don't understand quantities or concepts like biggest verse smallest. They may not understand that the numeral 5 is the same as the word *five*. Individuals with dyscalculia also have trouble with the mechanics of doing mathematics, such as being able to recall basic mathematics facts. They may understand the logic behind maths but not know how or when to apply what they know to solve math problems.

They also often struggle with working memory. For example, they may have a hard time holding numbers in mind while doing math problems with multiple steps.

Dyscalculia can be quantitative, which is a difficulty in calculating; qualitative, which is a difficulty in conceptualising mathematics processes or mixed, which is the inability to integrate quantity and space.

This can be summarised in the following way:

The language of mathematics: this is particularly true when the same words carry different meanings such as area or volume. If you follow the advice from the earlier section discussing EAL leaners and mathematics your dyslexic and dyscalculic students will be supported too.

Sequencing: much of mathematics relies on sequencing for success. Dyslexic learners can be quick to see the big picture but may sometimes make errors in following a sequence. One young woman with dyslexia Tony supported would often make errors when she turned over a page, writing one number at the bottom of one page and copying it incorrectly onto the next page.

Orientation: mathematical symbols can look very similar to each other for students with dyslexia. This is because of difficulties with orientation and direction. Indeed, an early sign of dyslexia might be children reversing letters and symbols.

Memory: some dyslexic students have difficulties memorising long strings of facts and figures or formulae and equations. It is worth remembering that, if Einstein managed a good mathematical career without memorising lots of facts, maybe we shouldn't be seeing this as important.

Confidence: all of these things can lead to students having a lack of confidence in their mathematical abilities. This can be overcome in the same way that you will ensure all learners come to believe in their mathematical abilities.

You should avoid asking a student with dyslexia to read questions aloud unless you are sure they will be comfortable doing this. By checking individual needs, you will know what specific support your dyslexic learners may need. It may be that coloured filters will support them in reading text or printing on a particular colour of paper. Being able to use highlighters to pick out key words is an important skill for all students and will support dyslexic learners across all subjects. Providing visualisation activities will help dyslexic learners; indeed they may excel in this area and be able to support other children in becoming more skilled at visualisation.

Other strategies that can be useful in supporting students with dyslexia and dyscalculia are:

The use of concrete objects and manipulatives.

Clear definitions of words and symbols available through displays.

The use of images of the numbers system such as number squares and number tracks.

The appropriate use of ICT.

As with much of the advice in this chapter, you will have noticed that all of this support is helpful for all learners.

Multicultural and anti-racist approaches

TAKING IT FURTHER: **FROM THE RESEARCH**

Tom Cooper, Annette Baturo and Elizabeth Warren explored three case studies in Queensland schools attempting to enhance the outcomes for students from indigenous backgrounds. It is the case that the performance of indigenous students lags two years behind that of non-indigenous students in Queensland schools. This is not helped by the fact that, in remote communities, teachers are generally inexperienced and only usually stay in a school for two years and do not have to gain the experience that would help them in working with their indigenous aides.

In the paper *Indigenous and non-indigenous teaching: Relationships in three mathematics classrooms in remote Queensland* (Available at http://eprints.qut.edu.au/3619/), they identified four key issues to be addressed:

Training: training in mathematics subject knowledge and mathematics pedagogy was needed for both the teachers and their teaching aides. It was also important to try both groups in developing effective partnerships.

Equality in partnerships: another training issue was to support teachers and aides in understanding, respecting and valuing each other's culture.

Communication: there was a lack of communication between teachers and aides in terms of deciding what should be taught and how it should be taught. This did not support aides in developing an understanding of mathematics pedagogy or meet the requirements of equality in partnerships.

The Westernised nature of the mathematics classrooms: all the mathematics presented was highly Westernised so one of the aims of the project became to develop indigenous contexts for the mathematics.

This offers us a good starting point for considering anti-racist approaches to learning and teaching mathematics. The mathematics that we teach in schools is not value or culture free. Multicultural and anti-racist practice does not assume that any other practice is overtly racist, rather it acknowledges that the historical and social context in which our current system of schooling has developed may advantage some groups over others. It also takes the view that these systems can be challenged in order to ensure access to mathematics learning for all learners. Values and cultural beliefs are embedded in:

- the content of the curriculum that we offer to students
- the ways in which we teach that content
- the environment in which children learn mathematics
- the view of mathematics we bring to the classroom as teachers
- the ways that we measure success in our mathematics classrooms.

So a multicultural and anti-racist approach to mathematics teaching would:

- include content that develops students' understanding of cultures other than their own through reflecting on cultural and linguistic diversity
- use resources that draw on students' cultural heritage, counter or challenge bias in materials such as textbooks or draw on students' own experience
- present positive images of learners as mathematicians and use familiar contexts as starting points, as well as illustrating the diverse cultural heritage of the discipline of mathematics. This may include photographs of your current learners being successful in mathematics or may be people from the communities you draw your students from who are successful in mathematics
- use teaching strategies that both develop student confidence with mathematical language and positive attitudes towards linguistic diversity. Children here will see linguistic diversity as a benefit to the classroom and would expect to hear a range of languages in the classroom
- encourage collaborative teaching and learning that offers challenge to the students, encourage learners to express and examine their own views, encourage learners to become involved in their own learning, encourage learners to pose their own problems
- develop anti-racist attitudes through mathematics by using data that critiques stereotyped views of particular groups of people. So for example looking at images in newspapers and carrying out a data handling activity may well show the images that newspapers present of minority ethnic groups in a small number of roles, as sportspeople or as underprivileged

- monitor achievement and grouping by ethnicity to ensure equal access to the curriculum.

This set of bullet points can be seen as a series of prompts for planning. It accepts the need for the numeracy strategy to form the basis for planning but offers a way in which key questions can inform the planning so that values of multiculturalism and anti-racism pervade the planning.

An activity that would support this view of teaching and learning uses a hundreds chart in a range of scripts or languages to develop children's understanding of place value and the Hindu–Arabic number system. (The Hindu–Arabic number system is the system widely used across the world, including Australian schools; it is also called the decimal number system or the base-ten number system. This system has developed from combining the Hindu and Arabic representations of number.)

In this activity give groups of children hundreds charts in a range of scripts that you have previously cut up to form a jigsaw. They have to reconstruct the hundreds chart – this will draw on and develop their understanding of the hundreds chart they are used to working with as they explore it with a new set of eyes.

১	২	৩	৪	৫	৬	৭	৮	৯	১০
১১	১২	১৩	১৪	১৫	১৬	১৭	১৮	১৯	২০
২১	২২	২৩	২৪	২৫	২৬	২৭	২৮	২৯	৩০
৩১	৩২	৩৩	৩৪	৩৫	৩৬	৩৭	৩৮	৩৯	৪০
৪১	৪২	৪৩	৪৪	৪৫	৪৬	৪৭	৪৮	৪৯	৫০
৫১	৫২	৫৩	৫৪	৫৫	৫৬	৫৭	৫৮	৫৯	৬০
৬১	৬২	৬৩	৬৪	৬৫	৬৬	৬৭	৬৮	৬৯	৭০
৭১	৭২	৭৩	৭৪	৭৫	৭৬	৭৭	৭৮	৭৯	৮০
৮১	৮২	৮৩	৮৪	৮৫	৮৬	৮৭	৮৮	৮৯	৯০
৯১	৯২	৯৩	৯৪	৯৫	৯৬	৯৭	৯৮	৯৯	১০০

Summary

This chapter took as its starting point the importance of supporting all learners in becoming as good at mathematics as possible. This is vital in order to maximise their life choices. Through drawing on a range of research studies and advice from organisations with the interests of children with specific learning needs at heart, you have learnt about a wide range of strategies you can use to support the learners in your class. The key areas we have focused on have been children with specific learning needs and children from minority ethnic groups, all of whom may not reach their potential if we do not adapt the way that we plan and teach.

Reflections on this chapter

As you read this chapter you may well have realised that most of the suggestions for changes and adaptations to mathematics teaching would benefit all learners in your classroom. By focusing on the needs of specific groups of children that you teach, you begin to plan for individual needs. And by doing this all learners will become more engaged in the mathematics you are teaching.

Tony has worked for many years with an educator and poet, Maresa MacKeith. She has cerebral palsy, is a wheelchair user and communicates through facilitated communication. That is, she slowly points at letters on a grid in front of her to form words that are spoken by a personal assistant. She has written powerfully about her experiences of schooling and describes her "special needs" of being fed, being washed, having to move from one place to another, as "ordinary needs not special needs." She sees the purpose of education very differently from many other people. She sees mathematics classrooms as places in which people learn to be with each other and value each other well as places in which we learn mathematics. We want to end this chapter with her vision for education.

> My vision is to have a world where we all know how to listen to each other and care for each other. To get there we need to grow up together . . . . My vision is that we create a system of learning that prioritises our relationships with each other rather than how we 'achieve' in competition with each other.

Taken from

MacKeith, M. (2012) 'Breaking the cycle of isolation and ignorance', in Cotton, T. (ed.) *Towards an Education for Social Justice: Ethics Applied to Education.* Oxford: Peter Lang.

Going further

Mathematics for dyslexics (including dyscalculia)

STEVE CHINN

Now in its third edition, as well as outlining the difficulties that learners with dyslexia or dyscalculia may have in the classroom it offers suggestions for you to support these students in overcoming these difficulties. The mixture of theory in the first part of the book together with practical examples in the second half of the book should help you in developing your practice to be more inclusive.

Chinn, S. and Ashcroft, R. (eds.) (2006) *Mathematics for Dyslexics (Including Dyscalculia)*. London: Wiley-Blackwell.

Overcoming difficulties with number: supporting dyscalculia and students who struggle with maths

RONIT BIRD

This practical book offers teaching plans for pupils aged 9–16 who have dyscalculia. The focus is on practical activities, visualisation and building self-esteem. A resource bank including games and other activities is provided on an accompanying CD.

Bird, R. (2009) *Overcoming Difficulties with Number: Supporting Dyscalculia and Students Who Struggle with Maths*. London: Sage Publishing.

Inclusive mathematics 11–18 (special needs in ordinary schools)

MIKE OLLERTON AND ANNE WATSON

This book is written by two innovative and very experienced teachers and shares the approaches that they have used to raise student achievement and opportunity with a specific focus on learners with special educational needs.

Ollerton, M. and Watson, A. (2005) *Inclusive Mathematics 11–18*. London: Continuum Press.

CHAPTER 13

ICT AND THE USE OF CALCULATORS IN TEACHING AND LEARNING MATHEMATICS

Introduction

As a teacher in the 21st century, it is vital that you can use your skills in ICT to support your teaching and that you are able to design opportunities for learners to develop their own ICT skills. You have probably noticed your own learning and facility to use information technology and digital platforms progress rapidly during this time. From your first time on a digital platform not being sure how the tools work to becoming confident in the ways in which the particular platform can be suited to best support the students. We will look at what we have learned from "teaching under lockdown" later in this chapter, but first we will offer an overview of tools that are available, describing how we can best select from the wealth of software out there, to make sure we are genuinely designing opportunities for students to develop their skills rather than just filling time working at a computer.

This chapter also discusses how you can use calculators to support student's learning, drawing on the research around the use of calculators in primary schools. The use of calculators remains a controversial area. The more we travel the more we realise that calculator use is one of those topics which engages politicians and parents alike.

The Australian National Curriculum uses the following diagram to describe the sort of skills that is expected students will develop during their time in school. The general capabilities are to be able to investigate, communicate and create using ICT. These skills are contained within more general skills of operating within social and ethical protocols and managing and operating, basically "being able to use" ICT.

DOI: 10.4324/9781003315155-13

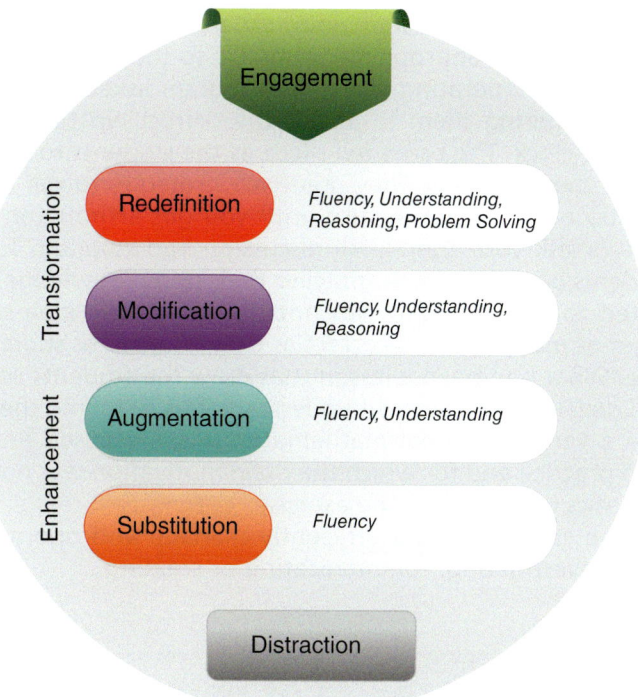

The accompanying statement in the mathematics section of the curriculum states:

> *students develop ICT capability when they investigate, create and communicate mathematical ideas and concepts using fast, automated, interactive and multimodal technologies. They use their ICT capability to perform calculations; draw graphs; collect, manage, analyse and interpret data; share and exchange information and ideas; and investigate and model concepts and relationships.*

Starting point

Tony had been observing an engaging lesson exploring a group of Year 2 students' understanding of time. They had been discussing times that they knew; the time they got up; the time they left for school; the time they went to bed; the time their mum got in from work and so on. They had recorded these times on clocks, both analogue and digital. This had led to some complex discussions about how to record times and why the big hand tends not to be where you would expect it to be. The lesson continued as the class moved into the IT suite containing enough computers for one between two students. The students all showed considerable skill in quickly logging on and finding the appropriate program, so they met the criteria of "managing and operating ICT." The program asked multiple choice questions requiring them to attach the correct written time to the time on the clock. The room fell silent as the students took it in turns to answer the questions. They used two strategies to get the correct answer; try each answer in turn until you got affirmation from the machine or ask your friend. All discussion had stopped. Tony asked the students how they were applying their prior knowledge, and they suggested to him that this was not the point. The idea seemed to be to get as many answers correct as they could in as short a time as possible. What had been a lesson that drew the students into exploring and developing their own understandings of telling the time had become a very traditional mathematics lesson focused around routine and practice and for which the main motivation was not learning mathematics but finishing the exercise.

So these students could efficiently manage and operate ICT but were not investigating, communicating or creating.

The appropriate use of personal computers and other devices

It is exciting that so many schools have facilities for using interactive software to support the teaching and learning of mathematics. However, we are concerned that sometimes when we see how the technology is used we find ourselves asking the question: "Is this software enhancing the learning of mathematics or is it simply replacing practices that might be more appropriately carried out by teachers?" To put this in the language of the curriculum, "Are we seeing students investigating, communicating and creating?"

So, how can we measure the effectiveness of a piece of software? We suggest there are some key criteria apply to any ICT you intend to use as a learning and teaching resource in your classroom. Ruben Puentedura has popularised the SAMR model to help teachers design and integrate digital learning experiences for their students. The model begins at the lowest level with *substitution*, which is really just a tool that is a direct substitute for something that already exists. At the next level, *augmentation* provides some functional improvement but still positions students as the consumers of technology. The model then moves to *modification* where the redesign of existing tasks occurs and finally to *redefinition*, which sees students as the makers, creating things that were previously inconceivable. Catherine Attard has adapted the SAMR model to point out that, without the pedagogy driving technology, its inclusion in the mathematics classroom can in fact be a hindrance to learning. At this lowest *distraction* level, she asserts that the digital technology in question should not be used.

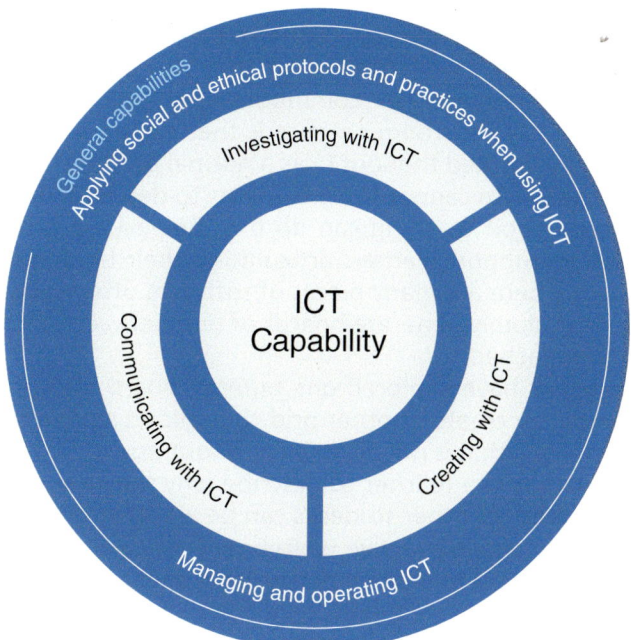

When we examine a piece of technology to see if it may improve the learning of mathematics, the first question might be, "To what extent does the resource provide an image or a representation of the mathematical idea I wish to teach?" In other words, does it model the mathematics that you are asking the students to engage with? An example of this that is seen in classrooms around the country would be the number line. This models the big idea of place value as well as supporting student in developing mental strategies for carrying out calculations. One example would be a zoom number line. Here the traditional number line has been developed to create a zoom number line. This number line allows the teacher to define the central number of the line and to alter the scale.

This capability of zooming in and out of a number line offers a view of the number system that is only available using ICT. It also meets the criteria of representing the number system and modelling the mathematics of place value. More than this, it allows us to begin discussions about such ideas as rational and irrational numbers. A rational number is one that can be placed exactly on a number line; an irrational number, such as pi, is a recurring decimal, and we do not know exactly where it is on the number line. We can also see that, no matter how far we zoom in on the number line, there will always be another zoom to be made. We can talk about infinity with primary mathematicians, and they love it. (There is a free version of such a number line at www.mathsisfun.com/numbers/number-line-zoom.html.) This sort of software allows the user to create their own number lines and so start to make decisions about which number line might be most helpful to solve a particular problem.

A second question could be, "To what extent does the resource encourage students to describe what they are doing?" This would ensure we are meeting the "communication" criteria. An example of this in everyday classroom practice is the teacher asking students, "How have you worked that out?" as a normal part of their teaching. Of course, a resource cannot force students to describe what they are thinking; only good teaching can do this, however, a program that can be used to support learners articulating their thinking is a good place to start. There are many pieces of software often called *number boards* that randomly generate boards of numbers according to criteria set by the teacher.

Once the class have explored one *number board* the software can then immediately create another grid that allows students to repeat processes in a way that is not available without using ICT. Students can be encouraged by the teacher to describe why they are making decisions at each step, or other students can be asked to give instructions to the learner at the interactive whiteboard, of course justifying why they are asking them to make a particular move.

TAKING IT FURTHER: **FROM THE CLASSROOM**

In *Mathematics Teaching 252*, published in June 2016 in the UK, Alf Coles explores the use of an interactive application called *Touchcounts*. This is a free, early number application, designed to make use of touch, gesture and sound to allow young learners to explore the world of counting. Alf has used this application in Early Years settings and suggests that it offers "rich possibilities for collaborative working between children, with or without adult intervention." He noticed children "calling numbers into existence" as a way of noticing the ordinal nature of counting and experiencing the cardinal aspect of number by creating "herds" of discs on their screen.

As with all applications, the best way to explore their effectiveness is through exploration with your learners. You can find more details about *Touchcounts* at www.mathsisfun.com/numbers/number-line-zoom.html.

A third question we can use to interrogate ICT resources is, "To what extent is the resource able to offer a range of representations of the same mathematical idea?" Those of you that use Excel or similar spreadsheet programs to support the study of statistics and data handling know that these programs can enhance the learning by allowing teacher and learner to move between different representations of the same data. They also allow students to compare quickly the representations and decide which they think most effectively represents the data. They have not had to invest time and effort on drawing the different representations and so will be happy to decide that a bar chart is more useful than a pie chart. If they had spent time drawing a pie chart it would have been difficult to motivate the same children to draw a bar chart in order to decide which is the most appropriate representation.

It is always exciting to see young learners motivated and excited by their developing understandings of mathematics. ICT has a vital role to play in supporting their learning and showing them that mathematics is something they can do and something that has an internal logic that they can explore and ask questions of. Our aim here is to suggest that an important role for the teacher is to ensure that the software we use when working with our students genuinely enhances their learning. Who knows? You may even find yourself waking up in the middle of the night with some mathematical questions.

The use of calculators

Many years ago, Tony was asked to write a "comment" piece for *Micromath*, the journal of the Association of Teachers of Mathematics in the UK, which focuses on the use of technology to support the teaching and learning of mathematics. As he was writing the article, the then minister for education made a statement that suggested children should not be "encouraged" to use calculators until they were eight years old. But after another change in minister for education it was seen as important that children in the Early Years had calculators available to support their role-play areas, especially if these are shops or cafés and such like.

However, the worry that calculators somehow "inhibit" the learning of mathematics persists to the extent that the current programmes of study in England contain the following sentences.

> *Calculators should not be used as a substitute for good written and mental arithmetic. They should therefore only be introduced near the end of key stage 2 to support pupils' conceptual understanding and exploration of more complex number problems, if written and mental arithmetic are secure.*

The use of calculators remains controversial around the world and seems to be an area that often concerns parents.

Whilst we would all agree that calculators should not be used as a substitute for good written and mental arithmetic, it does not follow that they should not be introduced until students reach a particular age. In fact, during the course of writing this chapter, Michael's five-year-old son discovered how to use the constant function on a calculator to skip count. This led to him spending long periods of time counting by 2s and 10s and then 5s and 1,000s, which was accompanied with many moments of noticing patterns, asking questions and genuine excitement, as he shared what he discovered with another family member (if you don't know how the constant function on your calculator works, read the handbook – if you haven't got a calculator, buy one!).

The research is this area, particularly that associated with the Calculator Aware Number (CAN) project, shows that students who have had free access to calculators from Year 1 and have followed a calculator aware curriculum are "more liable to compute mentally and adopt powerful mental strategies" than students who have not had access to calculators until they get to the age of 11.

We would argue that teachers in primary schools and Early Years settings should have calculators available all the time for the children they teach, whatever their age. A teacher's role is to teach their students how to use calculators effectively and to notice when they are useful. For example, if you are teaching multiplication by 10 you can give students a range of numbers and ask them to multiply these by 10 using a calculator and write down what they notice. Try it for yourself with these numbers:

14

157

3.8

2051

18.7

483.1

15.85

72.876

95,231

By exploring multiplication by 10 in this way the students themselves will realise that you don't "add a zero" when multiplying by 10. This is a commonly held misconception. Suggesting that calculators should not be available is a bit like suggesting that students should not be able to use a ruler as it might stop them estimating lengths. Calculators are just one of the tools of the trade for mathematicians. Our job is to teach students how to use them well.

Let us think about what progression in using calculators might look like. How can we plan a curriculum to teach the effective use of calculators? We think such a curriculum would include:

- teaching students how to use the calculator effectively so they can decide when it is appropriate to use a calculator and when other methods would be more efficient
- using calculators when the focus is on solving a problem rather than the process of calculation
- using the calculator as a tool to explore number patterns
- using a calculator to consolidate the learning of number facts and calculation strategies.

These ideas can be developed into the following progression sequence for developing calculator use:

Foundations in using calculators

Initially students may see calculators at home or in out-of-school contexts. It is important that these are replicated in school. Provide a calculator in any role-play environment where it is sensible or realistic to have one. This allows young students to use calculators to support their creative play and to begin to explore how the keys on a calculator operate. Young children also enjoy using calculators to display numbers that are familiar to them, such as their age or their house number.

Beginning to use calculators

When students are learning to read and write numbers up to 20, they can use calculators to illustrate two-digit numbers. Keying in these numbers can support them in their early understanding of place value. A 6-year-old child wanted to use a calculator to show me 17 (their brother's age) and asked the teacher how to, "spell seventeen." They looked at the possibilities, 17 or 71, then by looking at a number line agreed it should be 17. We have seen six-and seven-year-old children to explore adding and subtracting ten using a calculator. The teacher wrote

19

13

18

11

16

and asked the students to use a calculator to subtract 10 from each number and asked them what they noticed. As mentioned earlier, we might show students how to use the constant function on a calculator to notice the patterns that are formed when skip counting.

An activity that uses a calculator to support the development of mental skills is shown here:

5	9	3	8	2	12	9	19	15	5

Draw the number track on your whiteboard and ask your students to work in pairs with one calculator between them. The aim is to move from one box to the next using a calculator so that the calculator display shows the appropriate number after an operation and a number have been keyed in. The students take it in turns and in this way will keep a check on each other.

Developing the use of calculators

As students come to understand partitioning, a calculator game that can support them is Wipeout. Ask one student to write a three- or four-digit number on a piece of paper, say 3,582. Their partner then has to remove the digits from the display one at a time using the calculator. So, one way to do this would be:

$$3{,}582 \;-\; 3{,}000 \;=\; 582$$
$$582 \;-\; 500 \;=\; 82$$
$$82 \;-\; 80 \;=\; 2$$
$$2 \;-\; 2 \;=\; 0$$

You can add challenges to this game, e.g., change the order in which the digits must be eliminated, ban the use of particular keys, try to eliminate two digits at once. We promise you that the children in your class will very quickly come up with interesting and challenging rules for the game.

At this stage students can be introduced to the following calculator skills and learn how to:

● clear the display before starting a calculation
● correct mistakes using the clear entry (CE) key
● carry out one- and two-step calculations involving all four operations
● interpret the display correctly, particularly in the context of money
● recognise negative numbers on the display.

As students become more confident in their mathematics, they will use the calculator for the following:

- estimating the likely size of answers and checking calculations. We suggest stopping students pressing the = button on the calculator and ask them to estimate the answer – they are then very excited if they are proved correct by the calculator
- carrying out measurement calculations and interpreting the answer. For example, if we add 1m 35cm to 2m 15cm I will get 2m 50cm, but the calculator will show 2.5 rather than 2.50
- solving problems involving fractions. To do this, students will need to recognise decimal equivalents of fractions. This in itself is a useful exercise to carry out using a calculator.

Portfolio task 14.1

Write the following fractions as decimals by using a calculator.

$$\frac{3}{7} \quad \frac{1}{2} \quad \frac{4}{5} \quad \frac{1}{4} \quad \frac{1}{5} \quad \frac{2}{10} \quad \frac{7}{10}$$

For example, if you want to find 3/5 as a decimal you key in $3 \div 5 =$.

You should see 0.6.

Convert $\frac{1}{5}, \frac{2}{5}, \frac{3}{5}, \frac{4}{5}, \frac{5}{5}$ to decimals.

Do the same with

$$\frac{1}{4}, \frac{2}{2}, \frac{3}{5}, \frac{4}{4}, \frac{5}{5}, \frac{6}{10}, \frac{7}{10}, \frac{8}{10}, \frac{9}{10}, \frac{10}{10}$$

What do you notice about your answers to fifths and tenths?

Now explore all the other fractions from $\frac{1}{3}$s to $\frac{1}{12}$s.

Describe the patterns you are seeing.

Can you see how this activity helped develop a sense of the value of fractions and would have been impossible without the use of calculators?

Extending the use of calculators

Your most confident students will be able to use a calculator to:

- solve multi-step problems
- recognise recurring decimals. If you try to change 1/3 into a decimal using a calculator you will get 0.33333333; similarly, sevenths and ninths give recurring decimals.

A calculator-aware curriculum

It is interesting to me that there is very little research into the impact that using calculators has on students' understanding of mathematics. The only major piece of research in the UK took place between 1986 and 1989. It was called the Calculator Aware Number (CAN) project. Initially the project was based in 15 schools, but by the end of the process hundreds of schools had enrolled. Perhaps more importantly, all the schools that started the project, led by Hilary Shuard from Homerton College, Cambridge University, remained with it to the end.

The philosophy of CAN was that "Students should be allowed to use calculators in the same way that adults use them: at their own choice, whenever they wish to do so." The key principles that underpinned this philosophy were:

- students should always have a calculator available, and the choice to use it should be the student's not the teacher's
- traditional paper and pencil methods for the four operations will not be taught
- there should be a teacher emphasis on practical investigational and cross-curricular mathematics
- students should engage in mathematical activities that involve a range of apparatus
- teachers should support students in developing confidence in talking about numbers using precise mathematical language
- mental strategies should be emphasised and sharing children's mental strategies encouraged
- teaching and learning "number" should occupy less than 50% of the time spent teaching and learning mathematics.

Research carried out into the long-term impact of the CAN project was reported by Kenneth Ruthven in *Research Papers in Education* 12 (3), 1997, in the paper "The long term influence of a calculator aware number curriculum on pupils' attitudes and attainments in the primary phase." This study compared the attitude and attainment of students in post-project schools matched with non-project schools. This used national tests at ages 7 and 11 to compare attainments. The study found that more students attained at high levels in post-project schools than in non-project schools at age seven, but this was also the case for low attainment. This suggests that there had been a greater differentiation of attainment in post-project schools than non-project schools. Teachers in post-project schools thought that this may be the result of their not offering enough support and structure to lower performing students carrying out open, unstructured activities.

By age 11 there were no differences in post-project and non-project schools in terms of either attainment or enjoyment. That is to say, both groups of students did equally well on tests, and when asked if they enjoyed mathematics the responses were the same whichever group the students belonged to.

Using ICT during lockdown. What did we learn?

We have all had to adapt our use of ICT to support learning whilst schools have been delivering curriculum via remote learning. In fact, you have probably found your ICT skills improving very rapidly as you responded to using a range of platforms to support the learning and teaching of mathematics remotely.

Jess remembers the surge in the number of digital resources that were shared during the first few months of the pandemic. Many colleagues of hers were overwhelmed initially, not knowing how to filter through the wide array of tools and apps. Given the shorter time frames teachers had with their students during remote learning though, it forced teachers to become more critical and selective about the resources they implemented, constantly referring back to the question "How does this [program/platform/app/tool] enhance the learning?" a question we should always been asking ourselves as teachers.

During the various periods of lockdown Tony, Jess and Michael worked with nine other teachers from around the world to share experiences of working remotely. Whilst not all the teachers were focused on learning and teaching mathematics, we were all, from Brazil to Belize, from Myanmar to Mongolia, from Kenya to Australia, exploring how we were surviving teaching remotely and thinking carefully about what we hoped schools would look like on the return. We all agreed that we hoped the return to school would not mean a return to learning and teaching exactly as it was pre pandemic. We all hoped that schools and educators world-wide would be able to take this opportunity to transform education in general and, for us, as writers of this book, transform mathematics teaching.

So, we focused on what had changed for us during teaching under lockdown. We realised that initially many teachers and schools tried to replicate lessons and the curriculum exactly as it would have been if students had been in school. This didn't work, so many schools took the decision to reduce the amount of content they were trying to cover and to reduce teacher input placing trust on the students to decide when and where it would be best for them to learn. Maybe this offers lessons for us on the return to school. Maybe part of the transformation is to offer students more say in the "what and where" they learn. We also realised how important the social side of learning is because this is what was lost. A reminder that social connections are at the heart of learning. This reminds us that we need to make time for us to develop our social connections with each other and to develop a community of learners in our classroom.

We shared the experience of providing learning experiences at home, often in very short time frames. The parents of our students had all been asked to support their children in accessing learning at home. During the pandemic we realised that we cannot teach and our students cannot learn without the support of the community. This is something to cherish when we return to school and something we can see as supporting both our own and our students' feeling of connectedness to the communities we serve. As a colleague in Australia said, "We need to continue to consider everybody's role in education, particularly parents as partners in education." Students need stimulation from parents as well as from teachers. One of the questions that we all need to ask ourselves is, "How do we engage with our community to convince them that they are partners in their children's learning?"

Let us finish this section by posing a question to you, the reader.

What did you learn about learning and teaching mathematics whilst teaching (or learning) during the pandemic? How will you draw on this learning to develop the learning and teaching of mathematics in your classroom?

Summary

This chapter has illustrated how calculators and digital devices can be effective tools for teaching and learning mathematics. We have offered questions to help you in deciding which particular pieces of software will be useful to support your teaching and to enhance your students' learning.

We have aimed to give you the confidence to encourage children to use calculators at all stages of their learning, showing how calculators can be used progressively to develop mathematical skills appropriate to the age of learners so that they can develop both appropriate mathematical skills and appropriate calculator skills. In this way you have seen how the calculator is both a tool for learning mathematics and a tool for carrying out mathematics.

Finally, we shared recent experiences of using ICT to teach remotely and reflect on what this period has taught us about learning and teaching mathematics.

Reflections on this chapter

Mark kept coming up to Tony's desk and swapping a ruler that he was using for a new one. After he had repeated this four or five times – Tony was an over-patient teacher sometimes – he asked him what he was doing. He said he was trying to find a ruler that worked. Tony was confused so asked him what he meant. Mark said that he was measuring a line, he knew the answer was 8cm but all of these rulers kept giving him, "8 and a bit." When Tony went over to see what he was doing, he was using the end of the ruler as his starting point rather than the "0." Once Tony had shown him this, he didn't see the need to keep changing rulers!

It strikes us that those who suggest that calculators might inhibit children's learning of mental methods or slow down their learning of written calculation are trying to solve Mark's problem by taking away the ruler rather than showing him how to use it. A calculator cannot inhibit children's learning – the way that we teach children how to use it can, but then that is within our control. We hope that this chapter has offered you a way to view calculators and computers as powerful mathematical tools that we are in control of and maybe even encouraged you to get your old calculator out of your bag and explore how you can best use it both to teach your learners and to carry out mathematics yourself.

Going further

Technology enabled mathematics education: optimising student engagement

CATHERINE ATTARD

This book explores how teachers of mathematics are using digital technologies to enhance student engagement in classrooms, from the Early Years through to the senior years of school. The research underpinning this book is grounded in real classrooms. The chapters offer ten case studies of mathematics teachers who have become exemplary users of technology. Each case study includes the voices of leaders, teachers and their students, providing insights into their practices, beliefs and perceptions of mathematics and technology-enabled teaching. These insights inform a theoretical model, the Technology Integration Pyramid, for guiding teachers and researchers as they endeavour to understand the complexities involved in planning for effective teaching with technology.

Attard, C. and Holmes, K. (2020). *Technology Enabled mathematics Education: Optimising Student Engagement.* London: Routledge.

Transforming teaching: global responses to teaching under the Covid-19 pandemic

LUCY COOKER, TONY COTTON AND HELEN TOFT

This book shares the successes and the problems that were solved by a diverse group of educators during the global pandemic. The shared stories from around the globe will help and inspire any teacher to develop skills to support blended learning in whatever teaching situation they find themselves.

Including lessons to be learned from kindergarten to university, this book introduces new ways of working and pedagogical approaches appropriate for developing global skills. It importantly focuses on teacher narratives to aid personal reflection and encourages readers to take responsibility for their own professional development. Each chapter prompts teachers to reflect and build on new skills developed through distance and blended learning, use of technology and new ways of relating to students.

Calculators in classrooms

PAUL SWAN

Calculators in Classrooms provides teachers with ideas and guidelines along with games and activities for using calculators sensibly to teach mathematics. It also contains detailed notes on how to use a calculator, a review of research on using calculators, a calculators skills checklist and a comprehensive section on sensible calculator use – including techniques for monitoring and checking calculations.

Available at https://drpaulswan.com.au/shop/calculators-in-classrooms/

INDEX